NTIERS IN ELECTRONIC TESTING

Consulting Editor

Vishwani D. Agrawal

s:

niques and Tools for Embedded Systems Reliability

P. Prinetto
-7589-8
mory Memory Testing
s
7255-4
p) Testing for Plug and Play Test Automation

7205-8
ing for System-on-a-Chip
Iyengar & Chandra
119-1
uilt-in Self-Test

050-0
nect Diagnosis
eung
14-6
Testing for Digital, Memory, and Mixed Signal VLSI Circuits
.D. Agrawal
91-8
Boundary-Scan: A Guide to the IEEE 1149.4 Test Standard

86-8
, Diagnosis and Measurement

8
LSI Circuits
eng
5-1
Case Studies in System Test and Diagnosis
R. Simpson
3-3
ing and Design Debugging
Cheng
-X
CMOS Analog and Digital Circuits

-5
orks: Logic Synthesis and Verification Using Testing Techniques

-8

Thadikaran
5
ategies

X
of High-Density Random-Access Memories
kraborty

cts, Faults and Yield Loss

FAULT INJECTION TECHNIQ
RELIABILITY EVALUATION

FRO

Books in the serie

Fault Injection Tech
Evaluation
A. Benso &
ISBN: 1-4020
High Performance Me
R. Dean Adam
ISBN: 1-4020-
SOC (System-on-a-Ch
K. Chakrabarty
ISBN: 1-4020-
Test Resource Partitio
K. Chakrabarty
ISBN: 1-4020-7
A Designers' Guide to B
C. Stroud
ISBN: 1-4020-7
Boundary-Scan Interco
J. de Sousa, P.C
ISBN: 0-7923-7
Essentials of Electronic
M.L. Bushnell, V
ISBN: 0-7923-79
Analog and Mixed-Signa
A. Osseiran
ISBN: 0-7923-86
Design for At-Speed Test
B. Nadeau-Dosti
ISBN: 0-79-8669
Delay Fault Testing for V
A. Krstic, K-T. C
ISBN: 0-7923-829
Research Perspectives and
J.W. Sheppard, W
ISBN: 0-7923-826
Formal Equivalence Chec
S.-Y. Huang, K.-T.
ISBN: 0-7923-818
Defect Oriented Testing fo
M. Sachdev
ISBN: 0-7923-808
Reasoning in Boolean Netw
W. Kunz, D. Stoffe
ISBN: 0-7923-9921
Introduction to I_{DDQ} Testing
S. Chakravarty, P.J.
ISBN: 0-7923-9945
Multi-Chip Module Test St
Y. Zorian
ISBN: 0-7923-9920-
Testing and Testable Design
P. Mazumder, K. Ch
ISBN: 0-7923-9782-
From Contamination to Def
J.B. Khare, W. Maly
ISBN: 0-7923-9714-

FAULT INJECTION TECHNIQUES AND TOOLS FOR EMBEDDED SYSTEMS RELIABILITY EVALUATION

Edited by

ALFREDO BENSO
Politecnico di Torino, Italy

and

PAOLO PRINETTO
Politecnico di Torino, Italy

KLUWER ACADEMIC PUBLISHERS
BOSTON / DORDRECHT / LONDON

A C.I.P. Catalogue record for this book is available from the Library of Congress.

ISBN 1-4020-7589-8

Published by Kluwer Academic Publishers,
P.O. Box 17, 3300 AA Dordrecht, The Netherlands.

Sold and distributed in North, Central and South America
by Kluwer Academic Publishers,
101 Philip Drive, Norwell, MA 02061, U.S.A.

In all other countries, sold and distributed
by Kluwer Academic Publishers,
P.O. Box 322, 3300 AH Dordrecht, The Netherlands.

Printed on acid-free paper

Contents

Contributing Authors

Joakim Aidemark, Chalmers Univ. of Technology, Göteborg, Sweden
Jean Arlat, LAAS-CNRS, Toulouse, France
Andrea Baldini, Politecnico di Torino, Torino, Italy
Juan Carlos Baraza, Università Polytecnica de Valencia, Spain
Marco Bellato, INFN, Padova, Italy
Alfredo Benso, Politecnico di Torino, Torino, Italy
Sara Blanc, Università Polytecnica de Valencia, Spain
Jérome Boué, LAAS-CNRS, Toulouse, France
Joao Carreira, Critical Software SA, Coimbra, Portugal
Marco Ceschia, Università di Padova, Padova, Italy
Fulvio Corno, Politecnico di Torino, Torino, Italy
Diamantino Costa, Critical Software SA, Coimbra, Portugal
Yves Crouzet, LAAS-CNRS, Toulouse, France
Jean-Charles Fabre, LAAS-CNRS, Toulouse, France
Luis Entrena, Universitad Carlos III, Madrid, Spain
Peter Folkesson, Chalmers Univ. of Technology, Göteborg, Sweden
Daniel Gil, Università Polytecnica de Valencia, Spain
Pedro Joaquín Gil, Università Polytecnica de Valencia, Spain
Joaquín Gracia, Università Polytecnica de Valencia, Spain
Leonardo Impagliazzo, Ansaldo Segnalamento Ferroviario, Napoli, Itlay
Eric Jenn, LAAS-CNRS, Toulouse, France
Barry W. Johnson, University of Virginia, VA, USA
Johan Karlsson, Chalmers Univ. of Technology, Göteborg, Sweden
Celia Lopez, Universitad Carlos III, Madrid, Spain
Tomislav Lovric, TÜV InterTraffic GmbH, Köln, Germany
Henrique Madeira, University of Coimbra,Portugal

Riccardo Mariani, Yogitech SpA, Pisa, Italy
Joakim Ohlsson, Chalmers Univ. of Technology, Göteborg, Sweden
Alessandro Paccagnella, Università di Padova, Padova, Italy
Fabiomassimo Poli, Ansaldo Segnalamento Ferroviario, Napoli, Itlay
Paolo Prinetto, Politecnico di Torino, Torino, Italy
Marcus Rimén, Chalmers Univ. of Technology, Göteborg, Sweden
Chantal Robach, LCIS-ESISAR, Valence, France
Manuel Rodríguez, LAAS-CNRS, Toulouse, France
Frédéric Salles, LAAS-CNRS, Toulouse, France
Mathieu Scholive, LCIS-ESISAR, Valence, France
Juan José Serrano, Università Polytecnica de Valencia, Spain
Joao Gabriel Silva, University of Coimbra,Portugal
Matteo Sonza Reorda, Politecnico di Torino, Torino, Italy
Giovanni Squillero, Politecnico di Torino, Torino, Italy
Yangyang Yu, Univ. of Virginia, VA, USA

Preface

The use of digital systems pervades all areas of our lives, from common house appliances such as microwave ovens and washing machines, to complex applications like automotive, transportations, and medical control systems. These digital systems provide higher productivity and greater flexibility, but it is also accepted that they cannot be fault-free. Some faults may be attributed to inaccuracy during the development, while others can stem from external causes such as production process defects or environmental stress. Moreover, as devices geometry decreases and clock frequencies increase, the incidence of transient errors increases, and consequently, the dependability of the systems decreases. High reliability is therefore a requirement for every digital system whose correct functionality is connected to human safety or economic investments.

In this context, the evaluation of the dependability of a system plays a critical role. Unlike performance, dependability cannot be evaluated using benchmark programs and standard test methodologies, but only observing the system behavior after the appearance of a fault. However, since the Mean-Time-Between-Failures (MTBF) in a dependable system can be of the order of years, the fault occurrence has to be artificially accelerated in order to analyze the system reaction to a fault, without waiting for its natural appearance.

Fault Injection emerged as a viable solution, and it has been deeply investigated and exploited by both academia and industry. Different techniques have been proposed and used to perform experiments. They can be grouped in *Hardware-implemented*, *Software-implemented*, and *Simulation-based* Fault Injection.

The process of setting up a Fault Injection environment requires different choices that can deeply influence the coherency and the meaningfulness of the final results. In this book we tried to collect some of the most significant contributions in the field of Fault Injection. The selection process has been very difficult, with the result that a lot of excellent works had to be left out. The criteria we used to select the contributing authors were based on the innovation of the proposed solution, on the historical significance of their work, and also on an effort to give the readers a global overview of the different problems and techniques that can be applied to setup a Fault Injection experiment.

The book is therefore organized in four different parts. The first part is more general, and motivates the use of Fault Injection techniques. The other three parts cover Hardware-based, Software-implemented, and Simulation-based Fault Injection techniques, respectively. In each of these parts three Fault Injection methodologies and related tools are presented and discussed. The last chapter of Part 4 discusses possible solutions to speed-up Simulation-based Fault Injection experiments, but the main guidelines highlighted in the chapter can be applicable to other Fault Injection techniques as well.

Alfredo Benso
alfredo.benso@polito.it

Paolo Prinetto
paolo.prinetto@polito.it

Acknowledgments

The editors would like to thank all the contributing authors for their patience in meeting our deadlines and requirements. We are also in debt with Giorgio Di Natale, Stefano Di Carlo and Chiara Bessone for their valuable help in the tricky task of preparing the camera ready of this book.

PART 1

A FIRST LOOK AT FAULT INJECTION

Chapter 1.1

FAULT INJECTION TECHNIQUES

A Perspective on the State of Research

Yangyang Yu and Barry W. Johnson

University of Virginia, VA, USA

1. INTRODUCTION

The low-cost high-performance microprocessors are easily obtained due to the current state of technology, but these processors usually cannot satisfy the requirements of the dependable computing. It is not easy to forget that the recent America Online blackout affecting six million users for two and half hours, which was caused by a component malfunction in the electrical system, and the maiden flight tragedy of the European Space Agency's Ariane 5 launcher, which was caused by a software problem. The failures of critical computer-driven systems have serious consequences, in terms of monetary loss and/or human sufferings. However, for decades it has been obvious that the Reliability, Availability, and Safety of computer systems cannot be obtained solely by the careful design, the quality assurance, or other fault avoidance techniques. Proper testing mechanisms must be applied to these systems in order to achieve certain dependability requirements.

To achieve the dependability of a system, three major concerns should be posed by the procedure of the computer system design:

1. Specifying the system dependability requirements: selecting the dependability requirements that have to be pursued in building the computer system, based on the known or assumed goals for the part of the world that is directly affected by the computer system;
2. Designing and implementing the computing system so as to achieve the dependability required. However, this step is hard to implement since the

A. Benso and P. Prinetto (eds.), Fault Injection Techniques and Tools for Embedded Systems Reliability Evaluation, 41-48.

system dependability cannot be satisfied simply from the careful design. Therefore, the third concern becomes the one that cannot be skipped.

3. Validating the system: gaining confidence that a certain dependability requirement has been attained. Some techniques, such as using the fault injection technique to test the designed product, can be used to help to achieve this goal.

Dependability is a term used for the general description of a system characteristic but not an attribute that can be expressed using a single metric. There are several metrics, which form the foundation of dependability, such as Reliability, Availability, Safety, MTTF, Coverage, and Fault Latency. These dependability-related metrics are often measured through the life testing. However, the time needed to obtain a statistically significant number of failures makes the life testing impractical for most dependable computers. In this chapter, fault injection techniques are thoroughly studied as a new and effective approach to testing and evaluating the systems with high dependability requirements.

1.1 The Metrics of Dependability

Several concerns of the dependability analysis have been defined to measure the different attributes of the dependable systems.

- **Definition 1: Dependability**, the property of a computer system such that reliance can justifiably be placed on the service it delivers [LAPR_92], which is a qualitative system attribute that is quantified through the following terminologies.
- **Definition 2: Reliability**, a conditional probability that the system will perform correctly throughout the interval [t0, t], given that the system was performing correctly at time t0 [JOHN_89], which concerns the continuity of service.
- **Definition 3: Availability**, a probability that a system is operating correctly and is available to perform its functions at the instant time t [JOHN_89], which concerns the system readiness for the usage.
- **Definition 4: Safety**, a probability that a system will either perform its functions correctly or will discontinue its functions in a manner that does not disrupt the operation of other system or compromise the safety of any people associated with the system [JOHN_89], which concerns the non-occurrence of the catastrophic consequences on the environment.

- **Definition 5: Mean Time To Failure (MTTF),** the expected time that a system will operate before the first failure occurs [JOHN_89], which concerns the occurrence of the first failure.
- **Definition 6: Coverage**, C is the conditional probability, given the existence of a fault, the system recovers [JOHN_89], which concerns the system's ability to detect/recover a fault.
- **Definition 7: Fault Latency**, Fault latency is the time between the occurrence of the fault and the occurrence of an error resulting from that fault [JOHN_89].
- **Definition 8: Maintainability**, a measure of the ease with which a system can be repaired, once it has failed [JOHN_89], which is related to the aptitude to undergo repairs and evolution.
- **Definition 9: Testability**, a means by which the existence and quality of certain attributes within a system are determined [JOHN_89], which concerns the validation and evaluation process of the system.

There are some other metrics, which are also used to measure the attributes of systems with dependability requirements [LITT_93], but they are not as widely used as the ones we just discussed:

- **Definition 10: Rate of Occurrence of Failures**, a measure of the times of failure occurrences during a unit time interval, which is appropriate in a system which actively controls some potentially dangerous process.
- **Definition 11: Probability of Failure on Demand (PFD)**, the probability of a system failing to respond on demand, which is suitable for a 'safety' system and only called upon to act when another system gets into a potential unsafe condition. An example is an emergency shut down system for a nuclear reactor.
- **Definition 12: Hazard Reduction Factor (HRF)**, $HRF = \frac{1}{PFD}$

1.2 Dependability Factors

It is well known that a system may not always perform the function it is intended to. The causes and consequences of deviations from the expected function of a system are called the factors to dependability:

- **Fault** is a physical defect, imperfection, or flaw that occurs within some hardware or software component. Examples of faults include shorts or opens in electrical circuits, or the divided-by-zero fault in a software program [JOHN_89].

- **Error** is a deviation from accuracy or correctness and is the manifestation of a fault [JOHN_89]. For example, the wrong voltage in a circuit is an error caused by an open or short circuit as well as a very big number as the result of the divided-by-zero is an error.
- **Failure** is the non-performance of some action that is due or expected [JOHN_89]. Such as, a valve cannot be turned off when the temperature reaches the threshold.

When a fault causes an incorrect change in a machine stage, an error occurs. Although a fault remains localized in the affected program or circuitry, multiple errors can originate from one fault site and propagate throughout the system. When the fault-tolerance mechanisms detect an error, they may initiate several actions to handle the fault and contains its errors. Otherwise, the system eventually malfunctions and a failure occurs.

Figure 1.1-1. Fault, Error, Failure chain

1.3 Fault Category

A fault, as a deviation in the computer hardware, the computer software, or the interfaces of the hardware and software from their intended functionalities, can arise during all phases of a computer system design process—specification, design, development, manufacturing, assembly, and installation throughout its entire operational life [CLAR_95]. System testing is the major approach to eliminate faults before a system is released to field. However, those faults that are unable to be removed can reduce the system dependability when they are embedded into the system and put into use.

1.3.1 Fault Space

Usually we use fault space, $\Im$, to describe a fault. $\Im$ is usually a multi-dimensional space whose dimensions can include the time of occurrence and the duration of the fault (when), the type of the value or form of faults (how), and the location of the fault within the system (where). We should note that value could be something as simple as an integer value or something much more complex that is state dependent. In general, the complete proving the sufficiency of the fault model used to generate $\Im$ is very difficult, if not impossible. The fault modeling of the applied processor is obviously the most problematic. It is more traditional to assume that the fault model is sufficient, justifying this assumption to the greatest extent possible with the experiment data, the historic data, or the results published in literatures.

The corresponding statistical distribution of a fault type plays a very important role during the fault injection process, since only the fault inserted into a typical fault location has the value to be researched. As we know, several standardized procedures can be used to estimate the failure rates of electronic components when the underlying distribution is exponential.

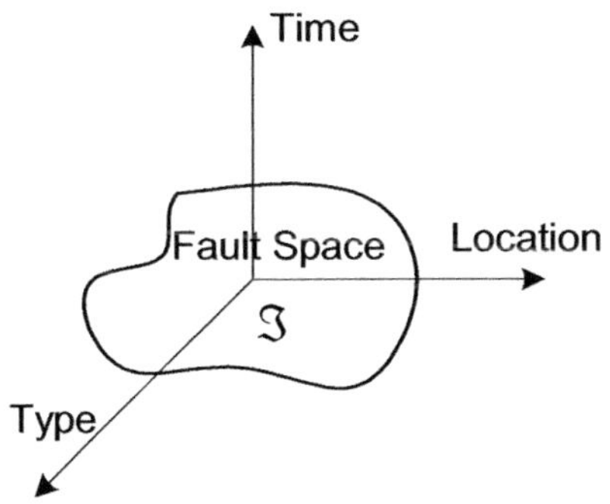

Figure 1.1-2. The System Fault Space

1.3.2 Hardware/Physical Fault

Hardware faults that arise during the system operation are best classified by their duration—permanent, transient, or intermittent. Permanent faults are caused by irreversible component damages, such as a semiconductor junction that has shorted out because of the thermal aging, the improper manufacture, or the misuse. Since it is possible that a chip in a network card that burns causes the card to stop working, the recovery can only be accomplished by replacing or repairing the damaged component or subsystem. Transient faults are triggered by environmental conditions such as the power-line fluctuation, the electromagnetic interference, or the radiation. These faults rarely do any long lasting damage to the component affected, although they can induce an erroneous state in the system for a short period of time. Studies have shown that transient faults occur far more often than permanent ones and are also much harder to detect. Finally, Intermittent faults, which are caused by the unstable hardware or varying hardware states, are either stay in the active state or the sleep state when a computer system is running. They can be repaired by replacement or redesign [CLAR_95].

Hardware faults of almost all types are easily injected by the devices available for the task. Dedicated hardware tools are available to flip bits on the instant at the pins of a chip, vary the power supply, or even bomb the system/chips with heavy-ions methods believed to cause faults close to real transient hardware faults. An increasingly popular software tool is a software-implemented hardware fault injector, which changes bits in processor registers or memory, in this way producing the same effects as transient hardware faults. All these techniques require that a system, or at least a prototype, actually be built in order to perform the fault testing.

1.3.3 Software Fault

Software faults are always the consequence of an incorrect design, at the specification or at the coding time. Every software engineer knows that a software product is bug free only until the next bug is found. Many of these faults are latent in the code and show up only during operations, especially under the heavy or unusual workloads and timing contexts.

Since software faults are a result of a bad design, it might be supposed that all software faults would be permanent. Interestingly, practice shows that despite their permanent nature, their behaviors are transient; that is, when a bad behavior of a system occurs, it cannot be observed again, even if a great care is taken to repeat the situation in which it occurred. Such behavior is commonly called a failure of the system. The subtleties of the system state may mask the fault, as when the bug is triggered by very particular timing relationships between several system components, or by some other rare and irreproducible situations. Curiously, most computer failures are blamed on either software faults or permanent hardware faults, to the exclusion of the transient and intermittent hardware types. Yet many studies show these types are much more frequent than permanent faults. The problem is that they are much harder to be tracked down.

During the whole process of software development, faults can be introduced in every design phase.

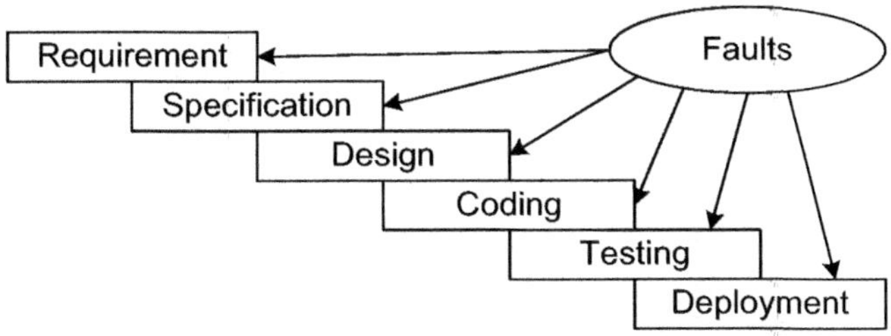

Figure 1.1-3. Faults Creation during the Software Development Process

Software faults can be cataloged into:

- **Function faults**: incorrect or missing implementation that requires a design change to correct the fault.
- **Algorithm faults**: incorrect or missing implementation that can be fixed without the need of design change, but this kind of faults requires a change of the algorithm.
- **Timing/serialization faults**: missing or incorrect serialization of shared resources. Certain mechanisms must be implemented to protect this kind of fault from happening, such as MUTEX used in operation systems
- **Checking faults**: missing or incorrect validation of data, incorrect loop, or incorrect conditional statement.
- **Assignment faults**: values assigned incorrectly or not assigned.

1.4 Statistical Fault Coverage Estimation

The fault tolerance coverage estimations obtained through fault injection experiments are estimates of the conditional probabilistic measure characterizing dependability. The term Coverage refers to the quantity mathematically defined as:

Equation 1:

$$C = P((\Pr operHandlingofFault) \mid (OccurrenceofAFault \in \Im))$$

The random event described by the predicate "proper handing of fault | occurrence of a fault $\in \Im$" can be associated with a binary variable Y, which assumes that value 1 when the predicate is true, and the value 0 when the predicate is false. The variable Y is then distributed like a Bernoulli distribution of parameter C, as evident from the definition of Coverage provided in the following equation and from the definition of the parameter of the Bernoulli distribution. It is well known that for a Bernoulli variable the parameter of the distribution equals the mean of the variable.

Equation 2:

$$C = E[Y] = 1 \cdot P(Y = 1) + 0 \cdot P(Y = 0) = P(Y = 1)$$
$$= \sum_{f \in \Im} P(Y = 1 \mid F = f) \cdot P(F = f)$$

If we assume that the fault tolerance mechanism will always either cover or not cover a certain fault f, the probability $P(Y = 1 \mid F = f)$ is either 0 or 1, and can be expressed as a function of the fault:

Equation 3:

$$y(f) = P(Y = 1 \mid F = f) = Y$$

Then, we are able to get the following expression for the Coverage:

Equation 4:

$$C = \sum_{f \in \Im} y(f) \cdot P(F = f)$$

The above equations depend on how accurately the constructed fault space $\Im$ describes the real fault space of the system. In many systems, the ability of the fault tolerance mechanism to tolerate a certain fault strongly depends on the operational profiles of the system at the moment of occurrence, As an example, software diagnostic routines that execute during a processor's idle time can be inhibited if the workload is very high: for such a system, the workload must be considered as one of the attributes of faults. The same concept extends to the other attributes as well, meaning that any attribute to which the fault tolerance mechanism is sensitive must be used as one of the dimension of the fault space $\Im$.

1.4.1 Forced Coverage

Suppose that we are able to force a certain distribution of the fault space. Then, we associate the event occurrence of a fault $\in \Im$ to a new random variable $\underline{F}$, different from F, which is distributed according to the new probability function $P(\underline{F} = f)$.

A new Bernoulli random variable $\underline{Y}$, different from Y, are introduced to describe the fault handling event, and the distribution parameter $\underline{C}$, different from C, of the variable $\underline{Y}$ is called the forced coverage and is calculated as follows:

Equation 5:

$$\underline{C} = E[\underline{Y}] = \sum_{f \in \Im} y(f) P(\underline{F} = f)$$

Although the two variable Y and $\underline{Y}$ have different distributions, they are related by the fact that the fault tolerance mechanism is the same for both, that is $Y = \underline{Y} = y(f)$ whether the fault f occurs with the fault distribution or with the forced distribution. Therefore, the values of C and $\underline{C}$ must also be related. In order to determine the relationship between the two distribution parameters, a new random variable P is introduced, which is also a function of the variable $\underline{F}$ and P is defined as

Equation 6:

$$P = p(f) = \frac{P(F = f)}{P(\underline{F} = f)}$$

When $\underline{F} = f$ occurs, the mean of P is calculated as

Equation 7:

$$E[P] = \sum_{f \in \Im} p(f) P(\underline{F} = f) = \sum_{f \in \Im} P(F = f) = 1$$

That is, when faults occur in the system with the forced distribution the ratio between the two probabilities, which the fault occurs according to the fault distribution and the forced distribution respectively, is equal, on the average, to unity.

The covariance ρ between the random variables $\underline{Y}$ and P, defined as the mean of the cross-products of the two variables with their means removed

Equation 8:

$$\rho = E[(\underline{Y} - E[\underline{Y}])(P - E[P])] = E[(\underline{Y} - \underline{C})(P - 1)] = E[\underline{Y}P] - E[\underline{Y}] - \underline{C}E[P] + \underline{C}$$
$$= \sum_{f \in \Im} y(f) p(f) P(\underline{F} = f) - \underline{C} - \underline{C} + \underline{C} = \sum_{f \in \Im} y(f) P(F = f) - \underline{C} = C - \underline{C}$$

The relationship between the forced coverage and the fault coverage of the system can therefore be expressed as

Equation 9:

$$C = \underline{C} + \rho$$

In order to better understand the meaning of the covariance ρ, the special case of the uniform forced distribution $P(\underline{F} = f)$ is considered

Equation 10:

$$P(\underline{F} = f) = \frac{1}{|\Im|} \forall (f \in \Im)$$

In this case, the value of the forced coverage $\underline{C}$ is just the total fraction of faults that are covered by the fault tolerance mechanism, called the Coverage Proportion, and indicated as $\dot{C}$

Equation 11:

$$\dot{C} = \underline{C} = E[\underline{Y}] = \frac{1}{|\Im|} \sum_{f \in \Im} y(f)$$

The random variable P is

Equation 12:

$$P = p(f) = |\Im| \cdot P(F = f)$$

The variable P expresses the ratio between the relative probability of the occurrence of the fault f and the same probability for the case when all faults are equally likely. The covariance between $\underline{Y}$ and P is then a measure of how fairly the fault tolerance mechanism behaves when it attempts to cover faults with higher or lower probability of occurrence than the average. An unfair fault tolerance mechanism that covers the most likely faults better than the least likely faults will lead to a positive covariance, while an unfair fault tolerance mechanism that covers the least likely faults better than the most likely faults will determine a negative covariance. If $\rho = 0$ then the fault tolerance mechanism is said to be fair.

In the more general case of a non-uniform forced distribution $P(\underline{F} = f)$, the covariance is a measure of how fairly faults occurring with probability $P(\underline{F} = f)$ will be tolerated as compared to faults occurring with probability $P(F = f)$. If $\rho = 0$, the fault tolerance mechanism is said to be equally fair for both distributions.

1.4.2 Fault Coverage Estimation with One-Sided Confidence Interval

The estimation of the fault coverage can be modeled as a binomial random variable X_i, where $X_i = \begin{cases} 1 \forall CoveredFault \\ 0 \forall Un\text{cov}erFault \end{cases}$

The fault injection experiments are performed to generate X_1, ……X_n, where each X_i is assumed to be independent and identically distributed. The expected value of the random variable is $E(X) = C$, and the variance of the random variable is $Var(X) = C(1\text{-}C)$.

Given the statistic

Equation 13:

$$S_n = \sum_{i=1}^{n} X_i,$$

the probability of $S_n = j$ is

Equation 14:

$$P(S_n = j) = \begin{bmatrix} n \\ j \end{bmatrix} C^j (1-C)^{n-j}, 0 \le j \le n,$$

If S_n out of n faults are observed to be covered, then C, the lower side of the confidence interval, satisfies the following equation

Equation 15:

$$P(S_n \ge s_n) = 1 - \gamma$$

where $P(S_n \ge s_n)$ is the probability S_n is greater than or equal to s_n, given that the Coverage value equals C_γ, and $\gamma$is the confidence coefficient. The signle-sided confidence interval is calculated as

Equation 16:

$$1-\gamma = \sum_{j=s_n}^{n} P(S_n = j) = \sum_{j=s_n}^{n} \begin{bmatrix} n \\ j \end{bmatrix} C^j (1-C)^{n-j}, 0 \le j \le n$$

Now, consider the case where all the faults are covered. In the case,

Equation 17:

$$1-\gamma = \sum_{j=s_n}^{n} \begin{bmatrix} n \\ j \end{bmatrix} C^j (1-C)^{n-j} = C_l^n$$

Rearranging the above equation and solving for n yields

Equation 18:

$$n = \frac{\ln(1-\gamma)}{\ln C_l}$$

1.4.3 Mean Time To Unsafe Failure (MTTUF) [SMIT_00]

After getting the coverage metric, we can define the other metric Mean Time To Unsafe Failure (MTTUF), which is the expected time that a system will operate before the first unsafe failure occurs. It can be proved that the MTTUF is able to be calculated

Equation 19:

$$MTTUF = \frac{MTTF}{(1 - C_{sys})} = \frac{1}{\lambda(1 - C_{sys})}$$

Where C_{sys} is the system steady state fault coverage and λ is the system constant failure rate.

2. AN OVERVIEW OF FAULT INJECTION

Fault Injection is defined as the dependability validation technique that is based on the realization of the controlled experiments where the observation of the system behavior in present of faults, is explicitly induced by the deliberate introduction (injection) of faults into the system.

Artificial faults are injected into the system and the resulting behavior is observed. This technique can speed up the occurrence and the propagation of faults into the system for observing the effects on the system performance. It can be performed on either simulations and models, or working prototypes or systems in the field. In this manner the weaknesses of interactions can be discovered, but this is a haphazard way to debugging the design faults in a system. It is better used to test the resilience of a fault tolerant system against known faults, and thereby measure the effectiveness of the fault tolerant measures.

Fault injection techniques can be used in both electronic hardware systems and software systems to measure the fault tolerance of such a system. For hardware, faults can be injected into the simulations of the system, as well as into implementation, both on a pin or external level and, recently, on an internal level of some chips. For software, faults can be injected into simulations of software systems, such as distributed systems, or into running software systems, at levels from the CPU registers to networks.

There are two major categories of fault injection techniques: execution-based and simulation-based. In the former, the system itself is deployed, and some mechanism is used to cause faults in the system, and its execution is then observed to determine the effects of the fault. These techniques are more useful for analyzing final designs, but are typically more difficult to modify afterwards. In the latter, a model of the system is developed and faults are introduced into that model. The model is then simulated to find the effects of the fault on the operation of the system. These methods are often slower to test, but easier to change.

From another point of view, the fault injection techniques can be grouped to invasive and non-invasive approaches. The problem with sufficiently complex systems, particularly time dependant ones, is that it may be impossible to remove the footprint of the testing mechanism from the behavior of the system, independent of the fault injected. For example, a real-time controller that ordinarily would meet a deadline for a particular task might miss it because of the extra latency induced by the fault injection mechanism. Invasive techniques are those which leave behind such a footprint during testing. Non-invasive techniques are able to mask their presence so as to have no effect on the system other than the faults they inject.

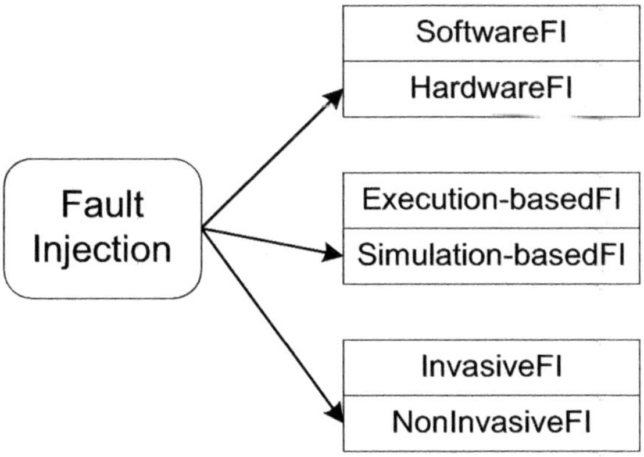

Figure 1.1-4. Fault Injection Technique Categories

2.1 The History of Fault Injection

The earliest work done for fault injection can be traced back to Harlan Mill (IBM)'s approach that surfaced as early as 1972. The original idea was to estimate the reliability based on an estimate of the number of remaining faults in a program. More precisely, it should be called *software fault injection* according to the current categorization of the fault injection techniques. This estimate could be derived from counting the number of "inserted" faults that were uncovered during the testing in addition to counting the number of "real" faults that were found during the testing.

Initially applied to the centralized systems, especially dedicated fault-tolerant computer architectures at early 70's, fault injection was used almost exclusively by industries for measuring the Coverage and Latency parameters of the high reliable systems. From the mid-1980s academia started actively using fault injection to conduct the experimental research.

Initial work concentrated on understanding the error propagation and analyzing the efficiency of new fault-detection mechanisms. Nowadays, fault injection has then addressed the distributed systems, and also, more recently, the Internet. Also, the various layers of a computer system ranging from hardware to software can be targeted by fault injection.

2.2 Sampling Process

The goal of fault injection experiments is to statistically evaluate as accurately and precisely as possible the fault coverage. Ideally one can calculate the fault coverage by observing the system during its normal operation for an infinite time, and calculating the limit of the ratio between the number of times that the fault coverage mechanism covers a fault and the total number of faults that occurred in the system.

For the purpose of statistical estimation, only a finite number of observations are made: from the statistical population Y, distributed as a Bernoulli random variable with unknown mean C, the Coverage, a sample of size *n* is selected, that is a collection of independent realizations of the random variable Y_k, indicated as $Y_1,\ldots,Y_n$. Assume that each realization of the random variable Y is a function of a fault that has occurred in the system, we get

$$Y_k = y(f_k)$$

Since it would take an extremely long time to observe enough occurrences of faults in the system, faults are instead sampled from the fault space $\Im$ and injected on purpose into the system. Indicating that $\underline{F}$ is the random variable associated with the event "fault $\underline{F} = f$ has been sample and injected", the sampling distribution is defined by the values of

$$P(\underline{F} = f).$$

Notice that the fault injection experiment forces the event " occurrence of a fault $\in \Im$ " in the system with the forced distribution. That is, sampling and injection a fault from the fault space with a certain sampling distribution is equivalent to forcing a fault distribution on the fault space. In most cases, it is assumed that sampling is performed with the replacement. When this is not true, it is assumed that the size of the sample is very small compared to the cardinality of the fault space. Therefore, Bernoulli analysis is still possible.

2.3 Fault Injection Environment [HSUE_97]

This fault injection environment is summarized in [HSUE_97]. It provides a good foundation of the fault injection environment, and the different fault injection applications may need to add their own components to meet their application requirements. It is very typical that a computer system that is under fault injection testing should have the components listed as follows:

- **Fault Injector** injects fault into the target system as it executes commands from the workload generator.
- **Fault Library** stores different fault types, fault locations, fault times, and appropriate hardware semantics or software structures.
- **Workload Generator** generates the workload for the target system as input.
- **Workload Library** stores sample workloads for the target system.
- **Controller** controls the experiment.
- **Monitor** tracks the execution of the commands and initiates data collection whenever necessary.
- **Data Collector** performs the online data collection.
- **Data Analyzer** performs the data processing and analysis

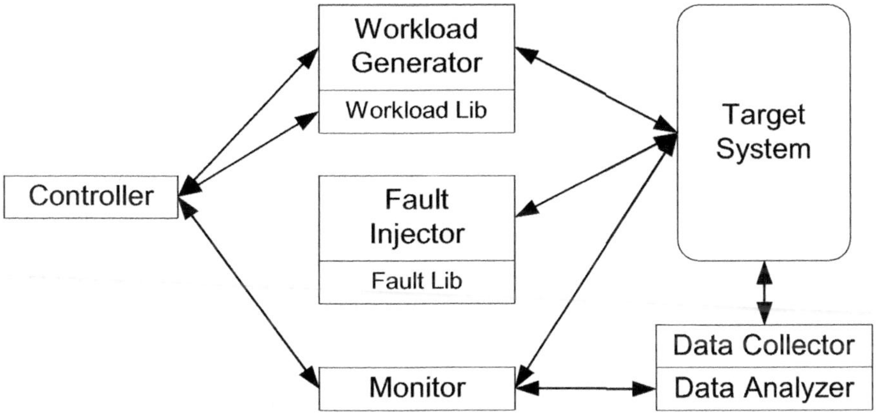

Figure 1.1-5. Basic Components of a Fault Injection Environment [HSUE_97]

2.4 Quantitative Safety Assessment Model

The assessment process begins with the development of a high-level analytical model. The purpose of this model is to provide a mathematical framework for calculating the estimate of the numerical safety specification. The single input to the model is the numerical safety specification along with the required confidence in the estimate. Analytical models include Markov models, Petri nets, and Fault Trees. For the purposes of a safety assessment, the analytical model is used to model, at a high level, the faulty behavior of the system under analysis. The model uses various critical parameters such as failure rates, fault latencies, and fault coverage to describe the faulty behavior of the system under analysis. Of all the parameters that typically appear in an analytical model used as a part of a safety assessment, the fault coverage is by far the most difficult to estimate.

The statistical model for Coverage estimation is used to estimate the fault coverage parameter. All of the statistical models are derived based on the fact that the fault coverage is a binomial random variable that can be estimated using a series of Bernoulli trials. At the beginning of the estimation process the statistical model is used to estimate the minimum number of trails required to demonstrate the desired level of the fault coverage. This information is crucial at the beginning of a safety assessment because it helps to determine the amount of resources required to estimate the fault coverage. One of the most important uses of the statistical model is to determine how many fault injection experiments are required in order to estimate the fault coverage at the desired confidence level.

The next step in the assessment process is the development of a fault model that accurately represents the types of faults that can occur within the system under analysis. The fault model is used to generate the fault space, $\Im$. Generally speaking, completely proving the sufficiency of the fault model used to generate $\Im$ is very difficult. It is more traditional to assume that the fault model is sufficient, justifying this assumption to the greatest extent possible with the experimental data, the historical data, or the results published in literature.

The input sequence applied to the system under test during the fault coverage estimation should be representative of the types of inputs the system will see when it is placed in service. The selection of input sequences is typically made using an operational profile, which is represented mathematically as a probability density function. The input sequences used for fault injection are selected randomly using the input probability density function. After a number of input sequences are generated then the fault coverage estimation moves to the next stage.

For each operational profile that is selected, a fault-free execution trace is created to support the fault list generation and analysis efforts later on in the assessment process. The fault free trace records all relevant system activities for a given operation profile. Typical information stored in trace is read/write activities associated with the processor registers, the address bus, the data bus, and the memory locations in the system under test. The purpose of the fault free execution trace is to determine the system activity such as memory locations used, instructions executed, and processor register usage.

The system activity can then be analyzed to accelerate the experimental data collection. Specifically, the system activity trace can be used to ensure that only faults that will produce an error are selected during the list generation process.

The set of experiments will not be an exhaustive set due to the size of the fault space, which is assumed to be infinite, but it is assumed that the set of experiments is a representative sample of the fault space to the extent of the fault model and to the confidence interval established by the number of samples. It is typically assumed that using a random selection process results in a sample that is representative of the overall population.

For each fault injection experiment that is performed, there are three possible outcomes, or events, that can occur as a results of injecting the fault into the system under analysis. First, the fault could be covered. A fault being covered means that the presence and activation of the fault has been correctly mitigated by the system. Here, correct, and likewise incorrect, mitigation are defined by the designers of the system as a consequence of the designers' definition of the valid and invalid inputs, outputs and state for the system. The second possible outcome for a given fault injection experiment is that the fault is uncovered. An uncovered fault is a fault that is present and active as with a covered fault and thus produces an error. However, there is no additional mechanism added to the system to respond to this fault, or the mechanism is somehow insufficient and cannot identify the incorrect system behavior. The final possible outcome for a given fault injection experiment is that the fault causes no response from the system.

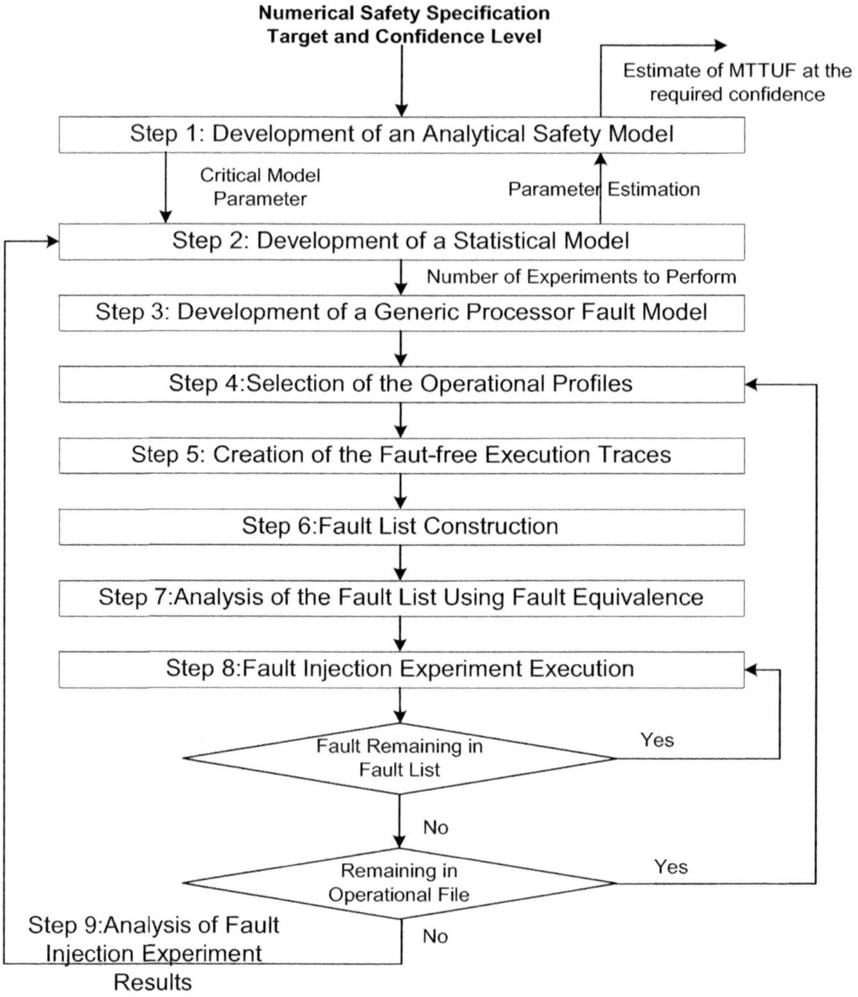

Figure 1.1-6. The Process for the Quantitative Safety Assessment

2.5 The FARM Model

The FARM model was developed in LAAS-CNRS in the early 90's. The major requirements and problem related to the development and application of a validation methodology based fault injection are presented through FARM model. When the fault injection technique is applied to a target system, the input domain corresponds to a set of faults *F*, which is described by a stochastic process whose parameters are characterized by probabilistic distributions, and a set of activation *A*, which consists of a set of test data patterns aimed at exercising the injected faults, and the output domain corresponds to a set of readouts *R*, and a set of derived measures *M,* which can be obtained only experimentally from a series of fault injection case studies. Together, the FARM sets constitute the major attributes that can be used to fully characterize the fault injection input domain and the output domain.

2.5.1 Levels of Abstraction of Fault Injection

Different types of the fault injection target systems and the different fault tolerant requirements for the target systems affect the level of abstraction for the fault injection models. In [ARLA_90], three major types of models are distinguished.

- **Axiomatic models**: using Reliability Block Diagrams, Fault Trees, Markov Chaining Modeling, or Stochastic Petri Nets to build analytical models to represent the structure and the dependability characteristics of the target system.
- **Empirical models**: using more complex and detailed system structural and behavioral information, such as the simulation approach and the operating system on top of the system, to build the model.
- **Physical models**: using the prototypes or the actually implemented hardware and software features to build the target system model.

In the case of the fault tolerant systems, the axiomatic models provide a means to account for the behavior of the fault tolerance system in the response to the fault. Although faults parameters can be obtained from the statistical analysis of the field data concerning the components of the system, the characterization of system behavior and the assignation of values to Coverage and execution parameters of the fault tolerance systems are much more difficult tasks since as we all know that these data are not a *priori* available and are specific to the system under investigations. Therefore, the experimental data gathered from empirical or physical models are needed to affirm or deny the hypotheses in selecting values for the parameters of the axiomatic model [ARLA_90].

2.5.2 The Fault Injection Attributes

The figure below shows that the global framework that characterize the application of fault injection for testing the FARM of a target system.

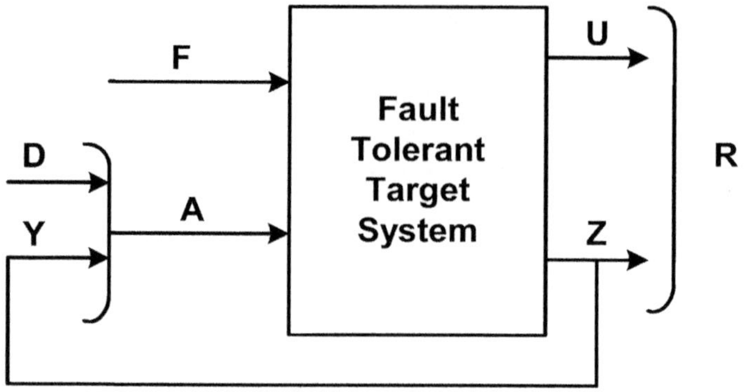

Figure 1.1-7. The Fault Injection Attributes

- **F**: the set of F of the injected faults; F is only a subset of all potential faults of the target fault tolerant system excluding fault tolerance deficient faults;
- **A**: the set of A that specifies the activation domain of the target system and thus of the injected faults as errors;
- **Z**: the set Z that defines the internal state of the target fault tolerant system;
- **U**: the set U that characterizes the functionality provided by the fault tolerant system to its users;
- **D**: the set D that designates the external input data;
- **Y**: the set Y that defines the set of current internal states;
- **R**: the set R of the readouts collected for each fault injection experiment to characterize the behavior in present of faults to account for larger subsets of both Z and U sets.
- **M**: the measure set M defines the experimental measures, such as latency estimate, fault dictionary entries, which are obtained by combining and processing the elements of the **FAR** sets.

The model of **FARM** describes explicitly the space of the injected faults in the input domain $I = \{D \times Y \times F\}$. It enables the target fault tolerant system to be described by means of a function *f* that relates the input domain I to the output domain $O = \{Z \times U\}$. The behavior of the target system is thus described by a sequence of states. The impact of a fault vector *f*(t) can be perceived when the fault is activated:

Equation 20:

$$\forall t, \exists d(t) and / or y(t) such that f(d, y, f; t) \neq f(d, y, f_0; t)$$

Where $f_0(t)$ is the vector "absence of fault".

This activation corresponds to the deviation from the expected trace:

1. Either as an internal error when only the state vector Z is altered:

   Equation 21:

   $$f(d, y_d, f; t) = (z', u; t) \neq (z, u; t)$$

   where z'(t) denotes an internal state distinct from the nominal one z(t),

2. Or as an error affecting the service when, as a result of a failure, the vector from U also deviates from the specified service:

   Equation 22:

   $$f(d, y, f : t) = \begin{cases} (z', u'; t) \neq (z, u; t) \\ (z, u'; t) \neq (z, u; t) \end{cases}$$

   where u'(t) denotes an output distinct from the nominal one u(t).

One thing that is worth mentioning here is that internal state set Z could be time related or non-time related. Indeed, the evaluation of a system does not depend at any time on all its internal states. This leads to make a partition of the state vector z(t) that distinguishes the state vector $z_s(t)$ – that characterizes the state variables that are sensitized at time t, that is, the internal variables that actually impact on the evolution of the system at time t – from the state vector z_d (t), which characterizes the variables that are not sensitized at time t. Such a distinction is useful in practice to account for the latent error:

Equation 23:

$$f(d, y_d, y_s f; t) = (z_d', z_s, u; t) \rightarrow f(d, y_d', y_s f_0; t) = (z_d', z_s, u; t)$$

This distinction is essential to characterize the fault-error-failure chain since dormant faults may not create erroneous behaviors and all erroneous states do not necessarily cause a failure, which has a direct impact on the controllability for the definition of the fault/error injection method to cover the F set and on the observability, in particular with respect to the control of the activation of the injected faults as errors and of the mutations induced by the propagation of these errors; and it is essential to design and implement the fault tolerant system since it is not necessary to try neither to observe nor to recover all system's states, which is especially important with respect to the observability of the response of the fault tolerant system in present of faults.

3. HARDWARE-BASED FAULT INJECTION

The hardware-based fault injection involves exercising a system under the analysis with specially designed test hardware to allow the injection of faults into the target system and to examine the effects. Traditionally, these faults are injected at the Integrated Circuit (IC) pin level, because these circuits are complex enough to warrant characterization through the fault injection rather than a performance range, and because these are the best understood basic faults in such circuits. Transistors are typically given stuck-at, bridging, or transient faults, and the results examined in the operation of the circuit. Usually these kinds of faults are injected using an intrusive approach.

Hardware fault injections occur in actual examples of the circuit after fabrication. The circuit is subjected to some sort of interference to produce the faults, and the resulting behavior is examined. So far, this has been done with both permanent and transient faults. A circuit could be attached to a testing equipment that operates it and examines the behavior after the fault is injected. This consumes time to prepare the circuit and test it, but such tests generally proceed faster than a simulation does. It is, rather obviously, used to test a circuit just before or in production. These simulations are non-intrusive, since they do not alter the behavior of the circuit other than to introduce the fault. Special circuitry should be included to cause or simulate faults in the finished circuit. These would most likely affect the timing or other characteristics of the circuit, and therefore be intrusive.

Hardware simulations typically occur in a high level abstraction of the circuit. This high level abstraction is turned into a transistor level abstraction of the circuit, and faults are injected into the circuit. Software simulation is most often used to detect the response to manufacturing defects. The system is then simulated to evaluate the response of the circuit to a particular fault. Since this is a simulation, a new fault can then be easily injected, and the simulation rerun to measure the response to the new fault. This consumes time to construct the model, insert the faults, and then simulate the circuit, but modifications in the circuit are easier to make than later in the design cycle. This sort of testing would be used to check a circuit early in the design cycle. These simulations are non-intrusive, since the simulation functions normally other than the introduction of the fault.

There are two types of fault injection techniques to catalog the hardware-based fault injection:

- **Forcing technique**: The fault is injected directly into an integrated circuit terminal, connector, etc, without any disconnection of parts. The fault injector probe forces a low or high logical level at the selected points.
- **Insertion technique**: A special device replaces a part of the circuit previously removed from its support. The device injects faults. The connection between two circuits is cut off before injecting the fault. Thus, the injection is performed on the side that remains at high impedance. Because there is not any signal forcing, there is not any danger of damage in the injected component.

3.1 Assumptions

1. The fault injector should have minor interference with the exercised system.
2. Faults should be injected at the locations internal to the ICs in the exercised system.
3. Faults that are injected into the system are representative of the actual faults that occur within the system. It means that both the random generated and the non-random generated faults can be injected into the system, and both the permanent and the transient faults can be injected into the system.

3.2 Advantages

1. Hardware fault injection technique can access to locations that is hard to be accessed by other means. For example, the Heavy-ion radiation method can inject fault into VLSI circuits at locations that are impossible to reach by other means.
2. This technique works well for the system, which needs a high time-resolution for the hardware triggering and monitoring.
3. Experimental evaluation by injection into actual hardware is in many cases the only practical way to estimate Coverage and Latency accurately.
4. This technique injects faults, which have low perturbation.
5. This technique is better suited for the low-level fault models.
6. Experiments are fast.

7. Experiments can be run in near real-time, allowing for the possibility of running a large number of fault injection experiments.
8. Running the fault injection experiments on the real hardware that is executing the real software has the advantage of including any design faults that might be present in the actual hardware and software design.
9. Fault injection experiments are performed using the same software that will run in the field.
10. No model development or validation required.

3.3 Disadvantages

1. Hardware fault injection can introduce high risk of damage for the injected system.
2. High level of device integration, multiple-chip hybrid circuit, and dense packaging technologies limit accessibility to injection.
3. Some hardware fault injection methods, such as the state mutation, require stopping and restarting the processor to inject a fault, it is not always effective for measuring latencies in the physical systems.
4. Low portability.
5. Limited set of injection points and limited set of injectable faults.
6. the setup time for each experiment might offset the time gained by being able to performed the experiments in near real-time.
7. Requires special-purpose hardware in order to perform the fault injection experiments.
8. .
9. Limited observability and controllability. At best, one would be able to corrupt the I/O pins of the processor and the internal processor registers.

3.4 Tools

- AFIT [Chapter 2.1]: a pin-level fault injection system developed by the Polytechnic University of Valencia (Spain);
- RIFLE [MADE_94]: a pin-level fault injection system developed at University of Coimbra, Portugal.

- FOCUS [GWAN_92]: a chip-level fault injection system developed at University of Illinois at Urbana-Champaign, U.S.A.
- FIST [GUNN_89]: a heavy-ion radiation fault injection system developed at Chalmers University of Technology, Sweden.
- MESSALINE [ARLA_90]: a pin-level fault forcing system developed at LAAS-CNRS, France.
- MARS [FUCH_96]: a time-triggered, fault-tolerant, distributed system developed at Technical University of Vienna, Austria.

4. SOFTWARE-BASED FAULT INJECTION

Nowadays software faults are probably the major cause of system outages. The fault injection technique is a possible approach to assessing the consequences of the hidden bugs. Traditionally, the software-based fault injection involves the modification of the software executing on the system under analysis in order to provide the capability to modify the system state according to the programmer's model view of the system. This is generally used on code that has communicative or cooperative functions so that there is enough interaction to make the fault injection useful. All sorts of faults may be injected, from the register and memory faults, to the dropped or replicated network packets, and to the erroneous error conditions and flags. These faults may be injected into the simulations of complex systems where the interactions are understood through the details of implementation, or they may be injected into the operating systems to examine the effects.

Software fault injections are more oriented towards implementation details, and can address the program states as well as the communication and the interactions. The software fault injector makes a system to run with faults so that its behavior can be examined. These simulations tend to take longer because they encapsulate all of the operations and details of a system, and they need more efforts to accurately capture the timing aspects of the system. This testing is performed to verify the system's reaction to the introduced faults and to catalog the faults successfully dealt with. This can be done later in the design cycle to show the performance for a final or near-final design. These simulations can be non-intrusive, especially if the timing is not a concern, but if the timing is at all involved the time required for the injection mechanism to inject the faults can disrupt the activity of the system, and cause the timing results that are not representative of the system without the fault injection mechanism deployed. All of these are caused by the injection mechanism, which runs on the same system as the software being tested.

4.1 Assumptions

1. Faults that are injected into the system should be representative of the actual faults that occur within the system.
2. The additional software required to inject the faults should not affect the functional behavior of the system in response to the injected fault.

Essentially, the assumptions state that the software that is used to inject the fault is independent of the rest of the system, and that any faults present in the fault injection software should not affect the system under analysis.

4.2 Advantages

1. This technique can be targeted to applications and operating systems, which are difficult to be done using the hardware fault injection.
2. Experiments can be run in near real-time, allowing for the possibility of running a large number of fault injection experiments.
3. Running the fault injection experiments on the real hardware that is executing the real software has the advantage of including any design faults that might be present in the actual hardware and software design.
4. No requirement for any special-purpose hardware; low implementation cost.
5. No requirement for the model development or validation.

4.3 Disadvantages

1. Limited set of injection instants: sometime at the assembly instruction level, only.
2. It cannot inject faults into locations that are inaccessible to software.
3. It does require a modification of the source code to support the fault injection, which means that the code that is executing during the fault experiment is not the same code that runs in the field.
4. Limited observability and controllability. At best, one would be able to corrupt the internal processor registers (as well as locations within the memory map) that are visible to the programmer, traditionally referred to as the programmer's model of the processor.
5. It is very difficult to model permanent faults.

6. The execution of the fault injection software could affect the scheduling of the system tasks in such a way as to cause hard, real-time deadlines to be missed, which violates assumption two.

4.4 Tools

- BOND [Chapter 3.1]: a software-based fault injection system for COTS applications, developed at Politecnico di Torino, Italy;
- XCEPTION [CARR_99_B] [Chapter 3.2]: a software implemented fault injection tool for dependability analysis developed at University of Coimbra, Portugal;
- MAFALDA [Chapter 3.3]: a fault injection environment for of Real Time COTS Microkernel-Based Systems, developed at LAAS-CNRS, Toulouse, France;
- DOCTOR [HANR_95]: an integrated software fault injection environment developed at University of Michigan, U.S.A.;
- EXFI [BENS_98_A]: a fault injection system for embedded microprocessor-based boards developed at Politecnico di Torino, Italy;
- FIAT [SEGA_88]: an automated real-time distributed accelerated fault injection environment developed at Carnegie Mellon University, U.S.A.

5. SIMULATION-BASED FAULT INJECTION

The simulation-based fault injection involves the construction of a simulation model of the system under analysis, including a detailed simulation model of the processor in use. It means that the errors or failures of the simulated system occur according to the predetermined distribution. The simulation models are usually developed using a hardware description language such as the Very high-speed integrated circuit Hardware Description Language (VHDL), or Verilog.

5.1 Assumptions

The simulated model has an accurate representation of the actual system under analysis.

5.2 Advantages

1. Simulated fault injection can support all system abstraction levels: axiomatic, empirical, and physical.
2. No intrusion is conducted in the real system.
3. Full control of both fault models and injection mechanisms.
4. Low cost computer automation, which does not require any special-purpose hardware.
5. It provides timely feedback for system design engineers.
6. Maximum amount of observability and controllability achieved.
7. Able to model both transient and permanent faults.
8. Allows modeling of the timing-related faults since the amount of the simulation time required to inject the fault is effectively zero.

5.3 Disadvantages

1. Large model development efforts needed.
2. Time consuming (the long experiment developing length).
3. Models are not readily available; rely on the model accuracy.
4. The accuracy of the results depends on the goodness of the model used.
5. Models may not include all of the design faults that may be presented in the real hardware.

5.4 Tools

- VFIT [Chapter 4.1]: a simulation-based fault injection toolset, developed at University of Valencia, Spain;
- MEFISTO-C [FOLK_98] [Chapter 4.2]: a tool to conduct fault injection experiments using VHDL simulation models, developed at Chalmers University of Technology, Sweden;
- ALIEN [Chapter 4.3]: a tool based on mutation techniques, developed at LCIS-ESISAR, Valence, France;
- VERIFY [SIEH_97]: a VHDL-based fault injection developed at University of Erlangen-Nurnberg, Germany;

6. HYBRID FAULT INJECTION

A hybrid approach combines two or more of the fault injection techniques to more fully exercise the system under analysis. For instance, performing the hardware-based or the software-based fault injection experiments can provide the significant benefit in terms of time to perform the fault injection experiments, it can reduce the initial amount of the setup time before beginning the experiments, and so forth. The hybrid approach combines the versatility of the software fault injection and the accuracy of the hardware monitoring. The hybrid approach is well suited for measuring extremely short latencies. However, given the significant gain in the controllability and the observability with a simulation-based approach, it might be useful to combine a simulation-based approach with one of the others in order to more fully exercise the system under analysis. For instance, most researchers and practitioners might choose to model a portion of the system under analysis, such as the Arithmetic and Logic Unit (ALU) within the microprocessor, at a very detailed level, and perform simulation-based fault injection experiments due to the fact that the internal nodes of an ALU are not accessible using a hardware-based or software-based approach.

6.1 Tools

- LIVE [AMEN_97] [Chapter 2.2]: a hybrid hardware/software fault injection environment developed at Ansaldo-Cris, Italy;
- A Software/Simulation-based Fault Injection environment [GUTH_95] developed at Chalmers University of Technology, Sweden.

7. OBJECTIVES OF FAULT INJECTION

Fault injection tries to determine whether the response of the system matches its specifications, in presence of a defined space of faults. Normally, faults are injected in the perfectly chosen system states and points, previously determined by an initial system analysis. The fault injector designers know the design in depth and so they design the test cases (type of faults, test points, injection time and state) based on a structural criteria and usually in a deterministic way.

Fault injection techniques provide a way for fault removal and fault forecasting which yields three benefits:

1. An understanding of the effects of the real faults and thus of the related behavior of the target system.
2. An assessment of the efficacy of the fault tolerance mechanisms embedded into the target system and thus a feedback provided for their enhancement and correction, such as removing designs faults in the fault tolerance mechanisms.
3. A forecasting of the faulty behavior of the target system, in particular encompassing a measurement of the coverage provided by the fault tolerance mechanisms.

In practice, fault removal and fault forecasting are not used separately; instead one is followed by the other. For instance, after rejecting a system by the fault forecasting testing, several fault removal tests could be applied. These new tests provide actions that can help the designer to improve the system. Then, it will be applied to another fault forecasting test, and so on.

7.1 Fault Removal [AVRE_92]

With respect to the fault removal objective, fault injection is explicitly aimed at reducing the presence of faults in the design/implementation of the fault tolerance system whose consequences would be deficiencies in their expected behavior when faced with faults they are explicitly intended to handle: faults are injected to uncover the potential fault tolerance deficiencies faults and thus, to determine the most appropriate actions to improve the fault tolerant systems. Fault removal involves a system verification to reduce the outcomes produced by the possible faults introduced in the design, development and prototype construction stages and also to identify the proper actions in order to improve the design.

Fault removal is composed of three major steps: verification, diagnosis, and correction. These steps are performed in an order: after it has been determined that the system does not match its specifications through verification the problem is diagnosed and, hopefully, corrected. The system must then be verified again to ensure that the correction succeeded. Static verification involves checking the system without actually running it. Formal verification is one form of static verification. Code Inspections is another. Dynamic verification involves checking the system while it is executing. The most common form of dynamic verification is testing. Exhaustive testing is typically impractical. Conformance testing checks whether the system satisfies its specification. Fault finding testing attempts to locate faults in the system. Functional testing tests that the system functions correctly without regard to implementation. Structural testing attempts to achieve path coverage to ensure that the system is implemented correctly. Fault-based testing is aimed at revealing the specific classes of faults. Criteria-based testing attempts to satisfy a goal such as the boundary value checking. Finally, the generation of test inputs may be deterministic or random. The above viewpoints may be combined. For example, the combination of fault finding, structural, and fault-based testing is called mutation testing when applied to software.

7.2 Fault Forecasting [ARLA_90]

In the case of fault forecasting, the major issue is to rate the efficiency of the operational behavior of the dependable systems. The type of test is mainly aimed at providing the estimates for the parameters that usually characterize the operational behavior of the error processing and the fault treatment, such as Coverage and Latency.

Fault forecasting aims to quantify the confidence that can be attributed to a system by estimating the number and the consequences of possible faults in the system. However, fault forecasting can be qualitative or quantitative. Qualitative forecasting is aimed at identifying, classifying and ordering the failure modes, or at identifying the event combinations leading to undesired events. Quantitative forecasting is aimed at evaluating, in probabilistic terms, some of the measures of dependability. There are two major approaches to the quantitative fault forecasting, which are aimed at deriving the probabilistic estimates of the dependability of the system: modeling and testing. The approaches towards modeling a system are different because the system can be considered in a stable state (that is, the systems level of reliability is "unchanging") or in a reliability growth state (that is, the reliability of the system is improving over time as faults are discovered and removed.). Evaluation of a system in stable reliability involves constructing a model of the system and then processing the model. Reliability growth models are aimed at performing reliability predictions from the data relative to past system failures.

8. FURTHER RESEARCHES

Fault injection is still a somewhat new technique, and there is still development being progressed to see what kinds of systems this can be applied to, and which systems it is appropriate to test in this manner. While there are some well understood mechanisms for injecting faults into certain kinds of systems, such as distributed systems, other systems still have basic techniques being designed, such as VLSI circuits. Often the method for inserting faults is very application specific, rather than generalized, and therefore the comparison of testing methods is difficult. Finally, even when results have been gathered, researchers are still uncertain as to what exactly the results mean, and how they should be used.

There are two difficulties that must be addressed before the use of fault injection can be fully applied:

1. The occurrence of no-response faults,
2. The large number of experiments that are needed to estimate the coverage with a reasonable confidence interval.

8.1 No-Response Faults

The first issue is to deal with the no-response faults that can occur. A no-response fault is one that is present, but it is not active. For an instant, a fault could be present in a given portion of memory that is never accessed by the system. Or if it is active, it is not producing an error due to the fact that its effect is being masked by the system. These faults reduce the efficiency of the assessment process because the fault injection experiments for these faults provide no useful information and yet require the maximum amount of resources. So, any fault injection experiments involving no-response faults can effectively be discarded when estimating the Coverage. Note that in order to discard the fault, it must be shown that the fault is truly a no-response fault. For an instance, a fault that occurs within a portion of memory that is never accessed by the system can be considered a no-response fault assuming a single fault occurrence so that the rest of the system is fault-free. In this case, the fault is discarded since it will always be a no-response fault. However, if the portion of memory where the fault occurs is used by the system, but not during the time interval for which the experiment is being conducted, then the fault cannot be considered a no-response fault. The reason for this is that the fault could eventually become active if the portion of memory where it occurs is accessed by the system. In this case, the fault is returned to the fault space as a potential candidate for selection when something such as the time of observation, input operational profile, and so forth changes that could activate the fault.

In addition, estimating Coverage for a dependable system typically requires a very large set of experiments in order to be able to generate enough significant events (that is, covered or uncovered faults) to estimate a coverage value with a reasonable confidence interval. So, reducing, or ideally eliminating, the number of fault injection experiments involving no-response faults would help to minimize the time and effort needed to perform the fault injection-based assessment, which is fairly resource-intensive in general.

8.2 Large Number of Fault Injection Experiments Required

The second issue related to the assessment of the dependable systems is to deal with the large number of fault experiments needed to estimate the Coverage. Even if the no-response faults can be eliminated using some sort of algorithmic processing to constrain the fault space, the number of fault injection experiments needed to estimate the Coverage with a reasonable confidence interval can still be extremely large. For instance, a rough rule of thumb is that a coverage estimate of 0.9_i, with a reasonable confidence interval (for example, 90%), requires performing approximately 10^i fault injection experiments. One can see how this could quickly become a huge liability, given that there will more than likely be limited time and resources to devote to performing the fault injection experiments.

Chapter 1.2

DEPENDABILITY EVALUATION METHODS

Classification

Tomislav Lovric

TÜV InterTraffic GmbH, Am Grauen Stein, D-51105 Köln - Germany

1. TYPES OF DEPENDABILITY EVALUATION METHODS

Dependability, the ability of a system to perform its specified function under permissible operating condition during a given time period, can be quantified using measures of reliability, availability, or time to failure. Safety, the absence of unacceptable risks, is a further measure of interest. Various methods for dependability evaluation exist, each with different variations and properties. In complex systems, more than one method is used. There is no one-for-all technique.

The best selection depends on the intended objective and other factors, e.g., the availability of tools, or the experience of the staff with a method. An operator may be interested in an estimate of the reliability of components to prepare a cost-effective maintenance plan. He could optimize stock size, availability, or down time. In terms of consequences or legislator requirements, his estimate doesn't need to fulfill precise confidence limits. On the other side a safety authority might require the system supplier to provide definitive evidence with given confidence limits that the safety is within the tolerable bounds. The authorities are not interested in best estimates for safety. Instead they need resistant worst-case conservative evidence of the system hazardous failure rate being lower than the tolerable limit in order to grant their authorization.

Nevertheless, before the eventual authorization they need also evidence that as far as reasonably practicable all has been done to build a safe system.

A. Benso and P. Prinetto (eds.), Fault Injection Techniques and Tools for Embedded Systems Reliabilit Evaluation 41-48.

Appropriate methods (including dependability evaluation methods) have to be applied all over the lifecycle of a system, from concept phase over design phases, to the eventual disposal phase of the system. Since the same techniques can be applied in different variations at different phases, a classification of methods is not directly obvious.

Here a classification is made by the intent of the method, whether it is used for forecasting or for verification of dependability. This leads to three classes of dependability evaluation methods that may provide useful dependability evidence (Table 1.2-1).

Table 1.2-1. Dependability Evaluation Classes

Dependability Evaluation by	
1. Analysis	Analysis of process (e.g., residual fault density in similar projects) and product (e.g., analysis of credible failures of different levels of abstraction of the system)
2. Field Experience	Evidence from observation (incident reports) for the used system from previous use in similar operational profiles
3. Testing	Providing representative inputs and testing for expected outputs for dependability testing equates to Fault Injection Testing (equipment failure testing)

It is noteworthy that field experience can be regarded as a specific kind of "long-term" testing. Testing and Field Experience are *measurements* of dependability properties, that providing *direct evidence*. Many practical fault injection campaigns that use an abstraction layer of the real system under evaluation are classified as "Analysis Methods" here. Specific methods and their applicability are discussed in the following sections.

2. DEPENDABILITY EVALUATION BY ANALYSIS

Analysis can be applied to various stages of the life cycle. The following considerations typically apply for safety evaluation, but similar applies to other dependability measures. In this case consequences of hazards include additional dependability categories (e.g., one train unavailable – minor delay ... all trains unavailable – complete loss of service).

Early in lifecycle, when only System Concept is available without realization details, typically some of the following measures are applied (without claim of completeness):

- Comparative Studies:
  Initial estimates based on experience from similar systems
- Preliminary Hazard Identification and Analysis PHIA:

Based on assumptions and functional system view; identifies initial hazards to be considered in the initial Hazard Log
- Risk Analysis and Risk Evaluation:
  Determines hazard frequencies/ consequences; determines hazard mitigation requirements (i.e. system safety requirements) depending on the acceptable risk
- System Hazard Analysis (SHA):
  When system description and mission profile is available; refines PHIA; this could add new hazards to the Hazard Log
- Subsystem Hazard Analysis:
  Refines SHA when System Architecture is available Depending on complexity of the system separate hazard analysis on different topics can be performed, e.g., separate Interface Hazard Analysis, or Operation and Support Hazard Analyses
- Consequence Analysis.

In the early phase the dependability evaluation methods are coarse. The derived dependability measure estimates are based on existing experience and expectations (often qualitative, not quantitative). They build goals that need to be reached by the system design. More detailed dependability evaluation methods are required when the system design becomes more detailed. If the detailed dependability evaluation does not support the initial goals, either the goals need to be revised (as far as acceptable), or the system design must be improved. This iterative process continues until the high level goals and the detailed evaluations and measurements approach each other.

Later in design phases, when architecture or system design details are available, the following measures are typically applied (without claim of completeness):
- Parts-Count Reliability Analysis:
  Any part failure rate (taken, for instance, from field records or MIL-HDB) adds to the system failure; simple method but mostly leads to underestimation of dependability (conservative estimates)
- Reliability Block Diagrams:
  Accounts for the architectural layout (serial/parallel) of the elements responsible for the reliability of the observed system function (e.g., a critical signal)
- Fault Tree Analysis (FTA):
  Top-down method, visually comprehensive, but events may be overseen. Assumptions are necessary, e.g., branches need to be independent to allow calculation of top event; Variations: Event Tree Analysis
- Failure Mode and Effect Analysis (FMEA) / Failure Mode, Effect and Criticality Analysis (FMECA) / Software Error Effect Analysis (SEEA):

Systematic bottom-up approach. Large amount of credible failure modes possible, possibly not completely known failure modes. Similar/related: Cause Consequence Diagrams; Common Cause Analysis

- Models/Simulations:
  Markov Models, Petri Nets, Finite State Machines, specific purpose, assumption validation required
- Usage Profile Analysis:
  Component unreliability might be tolerated by system architecture for a given usage profile
- Compliance / Comparison:
  By comparison of the product and process to other products and processes (e.g., development standards), dependability can be estimated. For example if the development process is found to be compliant to a set of qualitative requirements a "sufficient level" of reliability is assumed concerning failures caused by systematic faults. Estimated remaining software fault density is another example for evaluation by comparison
- Theory of Error Detecting Codes:
  This can be used for evaluation of reliability of systematically coded systems as the Coded Monoprocessor DIGISAFE
- Combinations of the above:
  Typically combinations and variations of the mentioned evaluation measures are used, e.g., different measures for different parts of the system,

Note that there is no strict distinction which measures are applied in which design phase. A coarse FTA can be performed on system level whereas a fine grained FTA might be performed on component level for a system with limited complexity. Variations also are possible, e.g., a FTA might be purely qualitative showing events and defenses, or it may be amended with failure rates allowing calculation of the undesired top event under specific assumptions, e.g., independence of branches.

A common denominator for all analysis methods is that analysis is performed on a more or less *abstract model* (refined within the lifecycle) of the system, and that it requires assumptions. Typically detailed information on low level components is used to calculate dependability measures on high level. Noteworthy here are the following reasons that could prevent analytic evaluation of dependability measures:

- Complexity too high: Analyzing a computer system as detailed state machine, or analyzing each credible failure mode is not feasible (by today), even if inner structure is completely known.
- Information unavailable: Detailed component failure modes or component inner structure are unknown (e.g., for integrated circuits).

- Modeling assumptions invalid: most analyses require independence of analyzed parts. If the necessary assumption cannot be confirmed the analysis is not credible. This occurs, for instance, when the investigated component cannot be further broken down into independent parts (e.g., consider a single computer with self-testing, test executions and computer hardware reliability are not independent).
- Derived results insufficient: To be credible the assumptions made during analysis must be valid. Therefore conservative assumptions must be chosen. E.g., use of credible and known failure rates or coverage values might be insufficient to make the required dependability target evident by analysis, even though it is reached in reality by the system.

3. DEPENDABILITY EVALUATION BY FIELD EXPERIENCE

Using field experience as evaluation method for dependability measures requires an existing system. Field experience can only be applied after a preceding operational phase of the system under investigation within a comparable application.

Dependability requirements may be so high that field experience cannot be used as sole method of dependability evaluation. To reach a confidence level of 1-γ that the hazardous failure rate is below λ by observing the system in the field without failure a test time T (observation time) of

$$T = -\ln(\gamma)*1/\lambda$$

is required. For a 95% confidence this means e.g., for a target of γ = 10e-9 hazardous failures per hour, that an observation time without failure is required of over 2 million years – usually not feasible even with multiple systems running in the field.

However, field experience can be used for specific properties, such as the evaluation of reliability of specific functions always used in a specific way (no input or state dependencies). Here a "proven in use" argument might apply, and replace the need for complete testing. Also for less demanding reliability targets or for availability measurements field experience might be sufficient, e.g., for λ = 10e-5 follows T = 34 years, which might be feasible (34 systems for one year). The increased confidence gained by a "working system" may be a further argument in an overall dependability evidence strategy even if given quantitative dependability targets cannot be proven alone by field experience.

Note that field experience is typically used not only to predict dependability for a new application, but also to validate predictions that were made during previous lifecycle phases of the system.

During the operational phase of a system, performance monitoring is used to maintain confidence in the dependability performance that were predicted before. High penalties for the supplier of a system might result from not performing per the specified dependability targets (especially availability targets).

A FRACAS (Failure Reporting And Corrective Action System) and a RAMS (Reliability, Availability, Maintainability, Safety) performance monitoring system is often an integral part of the operation and maintenance procedures (e.g., for railway systems required by [EN50126]). One method to measure reliability of a new system in the field is to run it in parallel and compare it to an already proven system (the golden system) that performs the same function.

4. DEPENDABILITY EVALUATION BY FAULT INJECTION TESTING

Systems were dependability measures need to be measured are usually Dependable Systems, i.e. systems that are build to cope with faults (i.e. using redundancy). Structural redundancy with independent elements typically can be mastered by analytic methods, whereas functional redundancy (e.g., diverse functions), time redundancy (e.g., multiple execution) and informational redundancy (e.g., double storage, coding) cannot. This is where direct evidence by testing could be a feasible way to provide the required evidence.

Testing is in any case required to verify the correct implementation of the fault handling mechanisms. However here the focus is on testing as measure for dependability evaluation, i.e., for evaluating the effectiveness of the implemented measures, not their correct implementation, which is assumed to be given here.

For dependability measures the test inputs are the faults that occur in representative operational scenarios. They are best measured in the field, but as discussed in the previous section the applicability of evaluation by field experience is restricted, and fault injection testing can be used as a sort of "time-lapse" between single fault occurrences.

Fault injection itself has evolved to a wide field by its own. Fault injection methods can be classified by the type of faults that are injected or by the implementation mechanisms. Fault injection can be done within an abstract simulation model (in this case is classified as analytic simulation method), or on the real system. It can be used to predict software reliability that remains after testing. Here, however, it is assumed that fault injection is used to estimate dependability measures where analysis is not feasible or would be overly complex. Mainly this means measuring the remaining hazardous failure rate caused by random hardware faults by injection of a representative set of hardware faults and comparing the system behavior against a golden run (without faults). The following aspects are crucial in order to obtain credible measurements:

- Representative fault model and fault selection (Faults):
  To be able to draw conclusions on the real system performance, the fault injection environment must be able to inject those faults that are representative for real faults using a representative distribution.
- Low intrusiveness (Activation):
  Intrusion introduced by the fault injection environment may corrupt the system behavior, i.e. delays caused by fault injection mechanisms might give diagnostic routines more time to detect a fault that in reality, or might even lead to the detection of the intrusion where the fault would not have been detected.
- Observability (Readouts):
  The fault injection environment must allow observation of all important information relevant to evaluate the system state and calculate measures from the experiments.
- Inspectability (Measurements):
  To convince a third party of the derived measures the complete experiments execution and evaluation must transparent and auditable (e.g., unequivocal storage of results, or repeatability of experiments).
- Realistic Environment:
  Besides the faults there are other system inputs that could influence the system behavior. They come from the system environment that must be realistic and representative (environment simulation).

5. CONCLUSION AND OUTLOOK

Analytical measures are applied first in a systems lifecycle (already in concept phase) and are refined as more detailed information on the system design becomes available. Analyses are widespread and used as far as possible, but application of analysis is prevented by missing detailed information, increased complexity, or non applicable assumptions. Alternatively to analysis, dependability evaluation by measurement can be applied. But evaluation of dependability measures in the field is practically restricted for demanding targets, e.g., high safety targets. This leads to fault injection as evaluation measure. But to obtain credible measurements from fault injection several requirements must be carefully considered and solved. Currently there is increasing interest in using fault injection for dependability evaluation as complement to other measures, but finding "Common Criteria" for this approach is left as important task for future.

Chapter 1.3

SOFT ERRORS ON DIGITAL COMPONENTS

An emerging reliability problem for new silicon technologies

Riccardo Mariani

YOGITECH SpA, via Lenin, 132/P, I-56017 San Martino Ulmiano (Pisa) - Italy

Abstract: Data from many application fields confirm that digital components implemented in modern Very Deep Sub Micron technologies are more and more susceptible to Soft Errors, i.e. bit-flip in memories or in sequential elements. It is an emerging reliability problem that cannot be furthermore ignored and requires actions to be carried out at the various steps of a component design and production flow.

Key words: Soft Errors, Single Event Upset, Radiation Effects, Signal Integrity, Crosstalk, Fault Tolerance, Safety Critical Systems

1. INTRODUCTION

When in the seventies sensitivity of circuits to radiation became for the first time an important topic of scientific journals [BIND_75] [MAY_ 79_A] [ZIEG_79] [GUEN_79] [MAY_ 79_B], only military and aerospace memory designers and nuclear physicists were probably really interested in that and started to investigate methods to survive the problem. This "cold" situation continued for a while, but in the middle of '90, someone started to be much more worried about effects of alpha particle and neutrons on new technologies: in last ten years, big electronics companies such as Intel, IBM, TI, increased a lot the investments in that field [ZIEG_96] [NORM_96] [CATA_99] [BORK_99] [JOHN_00] [RONE_01] [CATA_01] [GRAH_02] [BERN_02].

Moreover, in last years became clear that modern technologies would be more and more sensible not only due to increased disruptive effect of

A. Benso and P. Prinetto (eds.), Fault Injection Techniques and Tools for Embedded Systems Reliability Evaluation, 41-48.

particles in their smaller geometries, but also because sources of transient faults are increased: current system-on-a-chips (SOCs) built in 130 nm or smaller technologies contain many millions of gates and incredible lengths of interconnections, they work at very high frequency and with very small power supplies. Therefore they are plenty of effects such timing faults (cross-talk, ground bounce) and time-variant behavior of transistors.

Nowadays, "soft errors" i.e. an undesired behavior of a memory cell or a sequential element in a digital component due to transient faults, are officially considered one of the main causes of loss of reliability in modern digital components, as clearly stated in The International Technology Roadmap for Semiconductors 2001: "Below 100 nm, single-event upsets (soft errors) severely impact field-level product reliability, not only for memory, but for logic as well." and "The trend to greater integration of dynamic, asynchronous and AMS/RF circuits increases vulnerability to noise, crosstalk and soft error". And confirmed again also in the 2002 update of the Roadmap: "Since the operating voltage decreases 20% per technology node, increasing noise sensitivity is becoming a big issue in the design of functional devices (e.g., bits, transistors, gates) and products (such as DRAMs or MPUs). This is becoming more evident due to lower noise headroom especially in low-power devices, coupled interconnects, IR drop and ground bounce in the supply voltage, thermal impact on device off-currents and interconnect resistivities, mutual inductance, substrate coupling, single-event upset (alpha particle), and increased use of dynamic logic families. Consequently, modeling, analysis and estimation must be performed at all design levels.".

Therefore, manufacturers are starting to plan real action to reduce these potential effects also for terrestrial applications. Automotive is one of this application fields where soft errors are becoming more and more important. Today's vehicles host many microelectronics systems to improve efficiency, safety, performance and comfort, as well as information and entertainment. These systems are in general Electronic Computing Unit (ECU) based on 8, 16 and 32 bit CPUs with a considerable amount of memories, and they are interconnected using robust networks. These networks remove the need for the thousands of costly and unreliable wires and connectors used to make up a wiring loom. It is the "x-by-wire" revolution, that it transforming automotive components, once the domain of mechanic or hydraulic, into truly distributed electronic systems [BREZ_01] [LEEN_02] [BERG_02]. Of course such systems must be highly reliable, as failure will cause fatal accidents, and therefore soft errors must be taken into account.

How then to protect a circuit against soft-errors? Many techniques exist, firstly introduced for the design of memories: hardened technologies, intrinsic cells redundancy, data coding, etc... But we have to be ready to pay something for that: each applied protection technique will ask a price at least in term of longer design and verification time, smaller degree of re-usability, more area and lower performance. With the previous described increase of soft error rate also at terrestrial level, and with the relevance of effects not only for the memories but also for the glue logic, a significant extra cost will not be acceptable: designers should be able to insert as much fault-tolerance as really needed, not more.

In that scenario, a step ahead in the design tool chain is mandatory. Behavior verification of complex systems will not be enough: dependability validation will become a must. Therefore, fault-injection would be more and more important in the SOC design flow.

2. SOFT ERRORS

We define "soft" error the end effect of a transient fault causing a failure of a circuit where only data are destroyed (e.g., a bit-flip): we distinguish it from "hard" error where also the internal structure of the semiconductor material is damaged, and the fault is therefore permanent or at least intermittent. Transient faults can be caused by external (such alpha-particles hitting the silicon) or internal (timing faults due to cross talk, ground bounce) disruptive events, and generally manifest themselves as a transient pulse on the output of a logic cell.

If a transient fault is generated in a memory cell or in a sequential element (flip-flop, latch), a soft error could immediately occurs: otherwise the pulse should propagate itself through the logic network, and it will cause a soft error when latched by a memory cell or a sequential element.

Standard measure of reliability is MTTF (mean time to failure), but when soft errors are concerned, failure rates are generally expressed in FITs. A FIT is a "failure in time;" one FIT is a single failure in 1 billion (10^9) hours. As an example, a system that experiences 1 failure in 13,158 hours has a failure rate of 10^9 / 13,158 = 76,000 FITs. Concerning instead probability of faults, we usually speak of Soft Error Rate (SER).

2.1 Radiation Effects (SEU, SEE)

Most important sources of soft errors are radiation effects. In general, radiations impacting silicon devices can determine cumulative effects (such total ionization dose or displacement) or Single Event Effects (SEE), the last ones furthermore classified in Single Event Upsets (SEU) or catastrophic SEE such as Single Event Burnout (SEBO), Single Event Gate Rupture (SEGR) or SEL (Single Event Latchup).

Concerning SEU from a physic point of view, when an ion impacts a semiconductor device, it creates an ionization column (see Figure 1.3-1): in a region within some microns of the impacting node, within hundred of fs, an almost cylindrical track of e-h pairs is created. In presence of an electric field (depleted junction), e-h pairs are separated by drift and a "funnel" shaped potential distortion is generated: then, this additional charge is collected and a current spike (known as "prompt collection") can be observed. Then, in about a ns, the funnel collapses and diffusion effects occur (known as "delayed diffusion"): additional charge can be collected, smaller in magnitude but in a longer time scale. This phenomenon is usually represented with a triangular or a double-exponential spike [MESS_82].

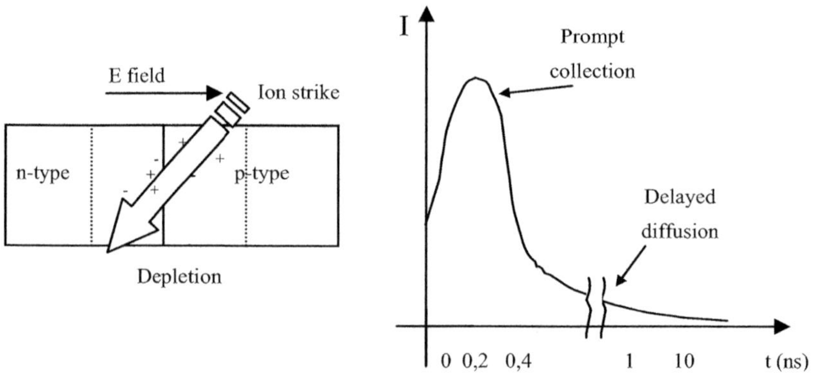

Figure 1.3-1. Current spike after a Single Event Upset

When this happens at the drain of one transistor in a flip-flop or memory, it can induce the change of logic state, i.e. a soft error. Of course that depends if the induced charge is enough to upset the stored data, i.e. if the collected charge Q_{coll} is greater than a critical charge Q_{crit}. Magnitude of Q_{coll} depends on many intrinsic factors such as size of the device, biasing, substrate structure, doping, and on event factors such as type of particle and in particular the charge of the particle, usually measured in LET (Linear Energy Transfer, e.g., an ion with a LET of 100 MeV-cm2/mg deposits approximately 1 pC of electronhole pairs along each micron of its track through silicon). Device robustness (i.e. Q_{crit}) mainly depends on design (layout, circuit architecture, etc...), and the more robust a design the greater its critical charge. As we will see in paragraph 2.3, technology scaling affects both quantities, i.e. both Q_{coll} and Q_{crit} are reduced: therefore SER effects must be carefully evaluated based on device features and design architectures.

SEU sources are mainly alpha particles, high energy and thermal neutrons from cosmic rays [BAUM_01]. Alpha particles are emitted by radioactive impurities such as ^{232}Th, ^{238}U or ^{210}Po, which are common in semiconductor materials and cause low but measurable emission rates in processed wafers, copper and aluminum metal layers, plastics, substrates and most significantly lead solders used in packaging the semiconductor devices. The concentration of these impurities can be reduced, but only at high cost. High-energy cosmic rays are present in space, where they threaten reliable operation of space missions and satellite electronics. As cosmic rays penetrate the atmosphere they interact in the upper layers and start cascades of secondary neutrons, protons and other sub-atomic particles. Some of these eventually reach ground level, cross a semiconductor device and, through nuclear interaction, may induce enough charge deposition to disrupt the logic state of a node.

High-energy neutrons (E>10MeV) are generally the dominant source of these disturbances at ground level, but also thermal neutrons can considerably contribute to SER, sometimes dominating it. This happens when they interact with ^{10}B, which constitute about 20% of the Boron commonly used as dopant in the manufacturing of semiconductor devices. In particular, the use of BPSG (BoroPhosphoSilicate Glass) dielectric layers can considerably increase the thermal neutron contribution to SER.

2.2 SER measurement and testing

The increasing relevance of SER problem convinced academics and companies to further investigate about methods and techniques to model and assess SEU sensitivity of devices, both during circuit design and testing.

During design, fault-injection is of course one of the base technique and it is widely discussed in this book. SEU-oriented tools also exist, such as SEMM used developed by IBM Labs [SRIN_94]. The Soft-Error Monte Carlo Model, or SEMM, is a tool just for modelling of SEU phenomenon. Its methodology is "predictive", since it is based on a physically based modelling approach without the need for arbitrary parameter fitting or high-energy beam testing. SEMM has been used to enhance reliability of many bipolar and CMOS ICs, thanks to the ability to estimate SER at an early stage of IC design.

After production, accelerated testing of silicon device is also used. Chips are irradiated with various hadrons such protons, neutrons, alpha particles and pions [HWAN_00] [HAZU_00_A]. In some cases, devices to be tested are designed in order to include special circuits to facilitate the SER measurement [HAZU_00_B]. Other technique is life-testing, i.e. using a tester containing hundreds of chips and evaluating their fail rate under nominal conditions. But it is very slow and very expensive, so it is not really used. Some alternatives either test more chips at the same time or they use different test conditions in a way to put the chips in a more sensitive state, e.g., at reduced operating voltage [ZIEG_00].

2.3 SEU and technology scaling

As previously described, technology scaling affects both the collected charge (reducing the collection volume) and the critical charge, minimizing capacitance, leakage, etc... Due to that, it is not easy to draw something similar to the Moore's law to strictly connect technology scaling and SER, and authors are discussing a lot on that.

In fact, different effects have been experimented at various technology steps (e.g., in 250 nm, SER seems to be dominated by thermal neutrons while in 180 nm atmospheric neutrons seem to be more relevant), and SER trends depends also on kind of circuit under evaluation (memories, logic, microprocessors, etc..). For example, in July 2001, Messer and co-workers of HP Labs identified latest generation CPUs having a 4,000 FIT i.e. the CPU experiences an SEU every 28 years at ground altitude [MESS_01] [CHEN_01]. They indicated that processor logic was responsible for half these failures, the other half coming from the large embedded cache memory. Ronen and others of Intel CPU R&D Labs indicated that SER increases by two times per generation. For a static latch, the induced noise scales with the supply voltage, but the current used to restore the logic is scaled by 0.7 every generation. Thus, the susceptibility increases by 43% per generation [RONE_01].

2.3.1 Trends in DRAMs, SRAMs and FLASHs

Concerning DRAMs, historically the worst technology in term of soft error sensitivity, it is almost agreed by all the sources that DRAM sensitivity has been slightly scaling down, since cell area scaling was more relevant than decrease of stored charge [ZIEG_00] [ZIEG_98] [HOFM_00] [MASS_96]. Moreover, advanced DRAM cell design such as trench or stack, maximize the capacitance while minimizing the junction collection volume. Finally, the operating voltage was not scaled as in other devices.

Concerning SRAMS, things are a bit contradictory. Some TI data report that elimination of BPSG in new 10Mb generations contribute to sharply decrease SER per bit, but the big growth of embedded SRAM brings to a global SER increase per chip. Therefore, FIT/bit is decreasing, but FIT/chip is increasing. The same appears in other references [PALA_01] [SCHE_00] [WROB_01].

Concerning FLASHs, SER sensitivity is generally very low, since the stored charge cannot be removed from the floating node directly by a radiation event through the dielectric layers that isolate the floating electrode. Some radiation effects have been reported in [NGUY_ 98]: however, it seems that for FLASH only external logic SER is a real concern.

2.3.2 Trends in Combinational Logic and Microprocessor

Combinational logic is in general much less susceptible to soft errors than memory elements. First of all, we define a soft error in combinational logic as the transient fault in a logic circuit that is subsequently stored in a memory cell or a sequential element of the circuit, causing a wrong value in that circuit. However, this transient fault might not be captured in a memory cell or in a sequential element because it could occur:

- a logical masking, i.e. the transient fault is generated in a portion of the combinational logic that is blocked from affecting the output due to other input values;
- an electrical masking, i.e. the transient fault is attenuated by subsequent logic gates;
- a latching-window masking, i.e. the transient fault reaches the memory cell or the sequential element, but not at the same time of the clock transition.

These masking effects led to a significantly lower rate of soft errors in combinational logic compared to storage circuits in equivalent device technology. However, due the fact that electrical masking is reduced by device scaling and due to higher clock rates, SER in combinational logic is dramatically increasing: in [SHIV_02] is estimated that passing from a 600 nm technology to 50 nm, SER in combinational logic will increase from 10^{-7} FIT to approximately 10^{2} FIT, or nine orders of magnitude!!

This of course strongly impacts the SER of microprocessors: several works by Compaq and Intel shows that even if SER sensitivity of caches and core logic decreased, logic errors are more and more dominant to the global SER [SEIF_01_A] [HOWA_01] [SEIF_01_B].

2.3.3 Trends in FPGA

During last five years, effects of soft errors in FPGAs (especially SRAM-based ones) have been carefully examined since FPGAs are now widely used in applications [MAVI_00]. A SRAM-based FPGA keeps its configuration by using SRAM cells; therefore they are sensitive to soft errors. For instance, following effects have been often experimented:

- a stress of the FPGA by creating a contention on two drivers of two internal cells, or a bus fight on internal tri-state busses;
- a change of functionality of the FPGA;

- a stress of the component surrounding the FPGA by changing the direction of a buffer.

To give an idea of the relevance of the problem, a typical 1 megagate SRAM-based FPGA (XCV1000) would have a ground-level firm error rate of 1200 FITs. Using this FIT rate:

- a complex system that uses 100 of these 1-megagate SRAM-based FPGAs would on average have a functional failure every 11 months;
- a system of similar complexity, if used in an aircraft or other high altitude environments, would have a SER that is 100-800 times worse, and a functional failure would occur every 12 to 36 hours;
- a product at ground level containing a single 1 megagate SRAM-based FPGA and shipped in 50,000 units, will be a significant risk of field failures: a failure every 17 hours!

2.4 Other sources of Soft Errors

In a terrestrial environment, other important sources of soft errors could be noise transients at the circuit, chip, and board or system level. These transients can be caused either by electromagnetic interference (EMI), i.e. coupling of high frequency EM signal, or by a combined effect of many issues such as crosstalk, ground-bounce, electromigration, hot carrier injection, negative biased temperature instability (NBTI) and process variation [CAIG_01].

Crosstalk gives the most important contribution to SER. A crosstalk induced error can occur when two or more parallel nets create a capacitor coupling between them. This cross-coupling capacitor can transmit a pulse from one net to another: when one of the two nets switches, this can cause the other net to temporarily assume a different logic than expected, in the same way of a SEU in combinational logic, and functional failures can occur if the glitch propagates and upsets a stored logic value in a following flip-flop or latch (see Figure 1.3-2).

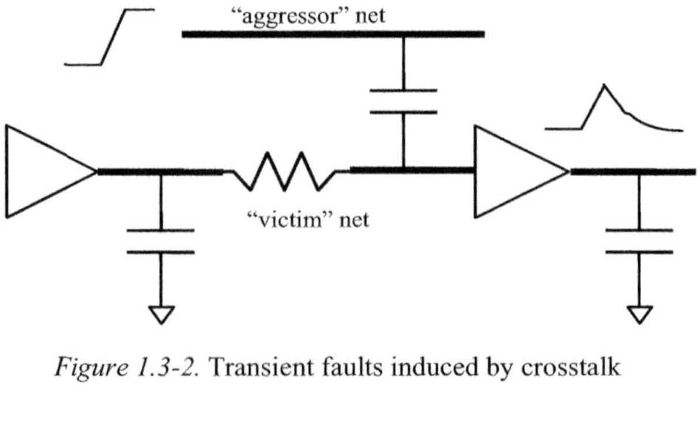

Figure 1.3-2. Transient faults induced by crosstalk

This phenomenon in principle is more deterministic than radiation-induced soft errors, since it depends on repeatable failures of certain logic operations. However, as the design shrinks, these problems are becoming more difficult to characterize, in particular the combined effect with other transient fault sources. Therefore, a certain kind of indeterminacy is present in standard cell modeling, affecting in particular timing analysis: and that generates somehow a new kind of noise that we could call "modeling noise".

3. PROTECTION AGAINST SOFT ERRORS

In principle, four methods are possible to achieve a dependable system: fault-avoidance, fault-removal, fault-tolerance and fault-evasion [LAPR_95]. Therefore, a combined use of following techniques would be needed to make a digital component robust against soft errors.

3.1 Soft Error avoidance

Soft error *avoidance* is mainly obtained from the use of radiation hardened technologies based on isolating substrates (SOI or SOS), often associated with an improved design of storage elements (see for instance [HARE_01] and [MUSSE_01]). In modern hardware design, fault-avoidance is also obtained by using methods to write, simulate and synthesize HDL descriptions in order to have clean and stable circuitry, i.e. without unknown states, with high testability coverage, etc.... In software design, fault-avoidance often implies formal methods, such as Model-based approach, algebraic approaches, Net-based approach, etc…. Also architectural level solutions, such as the use of asynchronous circuits, seem to contribute to reduce SER sensitivity [GOOD_89].

3.2 Soft Error removal and forecasting

Fault *removal* or design fault *forecasting* in the case of soft errors can be intended as the use of verification techniques (in primis, fault injection) able to assess during design phase if a given circuit can deliver its service in presence of faults. Designing components in new technologies would require in the EDA flow the use of verification of system dependability through the assessment of code and logic reliability by the injection of faults before and after the insertion of detection/correction. Fault injection is therefore the basic technique to design fault-tolerant circuits.

3.3 Soft Error tolerance and evasion

Since soft error avoidance through hardened technologies requires very high costs, in general soft error *tolerance* techniques are used. They are both software and hardware techniques and they can be introduced to perform different actions:

- fault confinement: soft error is contained before it can spread. This is generally done by using fault detection or consistency checks (such as multiple requests/confirmations) before performing a function;
- fault detection: to detect the soft error and to prevent it to generate bad data. Error Correction Codes (ECC, such as Hamming Codes), consistency checking, or protocol violation are used. In general, an arbitrary period of time may pass (latency) before detection occurs;
- fault masking: soft error is masked hiding the effects of the fault. It can be done using redundancy and majority voting (such as Triple-modular techniques);
- retry: since most problems are transient, a second attempt to execute the desired function can be done;
- diagnosis: to figure out what went wrong before to correction;
- reconfiguration: to replace or reconfigure a defective component, if the transient fault became permanent or intermittent;
- recovery: resume operation after reconfiguration in degraded mode;
- restart: re-initialize (warm or cold restart);
- repair: repair defective component;
- reintegration: after repair, go from degraded to full operation.

Run-time soft error *evasion* techniques imply a set of checks (either during start-up of the system or during normal system operations) in order to verify the system integrity respect to the golden behavior when the system it is not used in the application, in order to prevent failures. In general it includes Built In Self Test (BIST) or Built In Self Repair (BISR) for memories, Signature Analysis for CPUs, etc….

A detailed description of fault tolerant techniques is beyond the scope of this book: references and more detailed descriptions can be found for instance in [KANE_97] [WAKE_78] [HADJ_01].

3.4 SOC Soft Error tolerance

In the past (in particular in space or nuclear physics community) intrinsic redundancy was widely used as the main fault-tolerant technique, e.g., triple modular redundancy (TMR) of flip-flops, or internal coding using ECC. Nowadays, systems are enormously more complex than simply bunches of flip-flops: modern system-on-a-chips include million of gates, big CPU,

RAMs, etc...For these systems, main dependability problems concern memories and CPU. For the first ones, the most common solution is the use of various parity codes ranging from simple one-bit parity to single-error-correcting, double-error-detecting (SECDED) modified Hamming codes used for error detection and correction (EDAC) in large memory systems. For the CPUs, Full Dual Redundancy (FDR, two CPU running in parallel plus a comparator) or Mutual Dual Redundancy (MDR, two processors in a self-checking pair configuration) is mainly used in real applications.

However, modern CPU-based SOC cannot afford such very high costs: in fact any of these highly redundant techniques can double the circuit's size, which taking yield curves into account may triple IC product cost. In addition performance is significantly impaired because of slower system clocks, increased gate count and inclusion of redundant software. Some other solutions introduce redundancy at the software level, but in these cases the CPU performance is strongly affected by its fault-detection tasks.

In these cases, Dynamic Redundancy (i.e. only one processor, plus a fault detector to identify faulty behaviors of the processor) can be used: the cost of this approach is much less than FDR, MDR or TMR, but the fault detector must be carefully designed to have an acceptable detection coverage and a small latency.

4. CONCLUSIONS

We have seen that noise and radiation cause soft errors by disturbing the state of memory bits or registers inside semiconductor devices, and that Deep Sub-Micron CMOS device geometries of 180nm and below operating at Vdd of 1.5 volts or below are becoming highly susceptible to soft errors. On the other hand, SoCs contain many millions of flip-flops, not just in memory but also throughout the Register Transfer Logic (RTL), so the likelihood of soft error rates becoming noticeable increases with device complexity and is significant when a single device contains 50 million transistors (12.5 million gates) or more. Such devices are common in contemporary SoC employing multiple embedded processor cores, DSPs, IP blocks and on-chip SRAM.

In last ten years, Soft Error Rate problems in Integrated Circuits have been recognised by the space, nuclear and high energy physics communities and various techniques were used to simulate, design and verify circuits operating in radiation environments. However, these design and verification techniques were often doubling the circuit's cost, either increasing the design and verification time, or increasing area or decreasing performance. As we have seen in previous paragraphs, many research works and presentations at recent conferences indicate this issue is increasingly recognized and may become a barrier to progress for large SoCs fabricated on 100 nm DSM for commercial products.

Therefore a more sophisticated approach, both in design and verification, will be required for commercial applications where cost is the relevant factor. If HW/SW fault-tolerance techniques will be a possible answer from the design point of view, fault-injection will be mandatory for soft error sensitivity verification. This is confirmed by the fact that a very important and emerging norm concerning system robustness, the IEC 61508, gives precise requirements about the faults to be verified and tolerated to achieve the desired Safety Integrity Level (61508-1, tables A1-A13), and precise recommendation for the methodology to be used during system design (61508-2, table B-5). In particular it is stated that "Fault insertion testing (when required diagnostic coverage > 90%) is Highly Recommended and Mandatory" [LEEN_02].

PART 2

HARDWARE-IMPLEMENTED FAULT INJECTION

Chapter 2.1

PIN-LEVEL HARDWARE FAULT INJECTION TECHNIQUES

Pedro Gil, Sara Blanc and Juan José Serrano

Polytechnic University of Valencia,Fault Tolerant Systems Group (GSTF) - Spain

Abstract: Among hardware fault injection techniques involved in the validation of fault tolerant systems, pin-level fault injection is one of the more relevant techniques of the last years. The technique is suitable for injecting faults in the real system under test or in a prototype. Tools based on this technique are generic being possible the reusability of the tool. They are developed externally to the target causing no execution overhead on the system. This chapter presents a general overview about pin-level fault injection with a state of the art including the more remarkable features of the technique. Finally the tool *AFIT* developed by the Polytechnic University of Valencia (Spain), is described as a real implementation of the technique. Moreover, it presents an example about the methodology carried out in fault injection campaigns and the usability of the pin-level fault injection technique.

Key words: Dependability, pin-level fault injection, fault tolerance, hardware fault injection techniques.

1. INTRODUCTION

Pin-level fault injection is a hardware fault injection technique based on the idea of perturbing integrated circuits with faults introduced through the pins that emulate both external and internal faults. Physical fault models include open lines, bridges, and stuck-at faults. Faults are directly injected on the integrated circuit pins that belong to address or data busses, control lines, clock or oscillator inputs, peripheral inputs or outputs, etc. The fault duration is easily controlled as well as the number of pins or locations

A. Benso and P. Prinetto (eds.), Fault Injection Techniques and Tools for Embedded Systems Reliability Evaluation, 41-48.

injected. Both single and multiple faults can be injected perturbing only one pin of the integrated circuit or more than one at the same time.

The chapter is organized as follows: Sections 2 examines the state of the art of the pin-level fault injection technique. Section 3 analyses the FARM model. Section 4 is the description of a particular fault injection tool of one pin-level fault injection tool developed by the Fault Tolerant Systems Group of the Polytechnic University of Valencia, Spain and includes a practical case study. A critical analysis is provided in Section 5.

2. STATE OF THE ART

The technique is put into practice using external tools that need same level of synchronization with the target system under test. Among bibliography we find out pin-level fault injection tools before the 90's [ARLA_90], [KARL_95], [MADE_94], [MART_99]. Currently, the development of injection tools based on this technique is scarce. The increase of clock frequencies and the integration density of new circuits are the most relevant causes.

2.1 Fault injection methodology

Experimental validation can be done in many stages of the system life cycle. When a prototype or the real system is available, some validation techniques as pin-level fault injection can be used. Since the fault rates of integrated circuits during their operational phase is quite low, it is more efficient to stimulate the appearance of faults in the system and measure its consequences.

When validating a system by means of an experimental technique, especially if pin-level fault injection is used, we have to care for errors derived from the measurement procedures and also for those that the measurement equipment induces over the system. If we also take into account that nowadays systems are becoming very complex and fast, we soon realize the necessity of injecting faults of very short duration and the importance of the dismissal of the invalid experiments obtained during the injection campaign.

2.1.1 Fault injection

Prerequisites of an injection tool are the possibility of perturbing the target system with permanent or intermittent faults as well as transient faults of very short duration. The tool must be capable of performing unattended experiments (automation) and should also provide output signals to reset the system under evaluation. These signals have to drive the system to an initial state, providing the same conditions for every experiment in order to achieve the statistical correctness of the acquired data. Furthermore, the tool should provide some way of pointing out the beginning of the experiment and the effectiveness of the fault.

2.1.2 Data acquisition

When a fault is injected in a system, it is important to extract all possible information about the fault effect and the system behavior in the presence of the fault. Usually the needed time to inject a fault is very low as well as the readout acquisition time of the system reaction. However, to process all this information in order to extract measures is a high consuming task. If data acquisition is done in parallel within the injection and readouts are logged in a unique file, measurement could be done off-line. The log file should register also the experiment number, the fault location and duration, the type of fault and all possible timing information about the Error Detection Mechanisms (EDM) response.

2.1.3 Data processing

Once the injection campaign has been completed, the readout processing takes place. Glitches and ineffective faults should be discarded. The set of valid experiments is the set of every possible sequence of events that starts in a well-known state and returns to this state after visiting other states. Readouts must be parsed and if one experiment does not belong to the valid set of experiments it is assumed that an observation error has occurred and it is discarded. The parser can extract in parallel the timing information from the log file in order to calculate the error detection or system recovery latencies for every valid experiment.

2.2 Pin-level fault injection techniques and tools

There are two representative techniques (Figure 2.1-1):

- Socket insertion technique:
  A special device replaces a part of the circuit previously removed from its support (socket for an integrated circuit, bus connector, etc.). The device connects two parts of the circuit. The connection is cut off before injecting the fault and the injection is performed on the side that remains

at high impedance. The physical fault models are stuck-at, bridging, and open lines.

- Active probes technique (also called Forcing technique):
  The fault is injected directly on an integrated circuit terminal, connector, etc, without any part disconnection. It is more useful due to the high frequencies and packages (surface mount) of the modern integrated circuits. The fault injector probe forces a low or high logical level at the selected point. Usually, the physical fault models are stuck-at-(0,1)

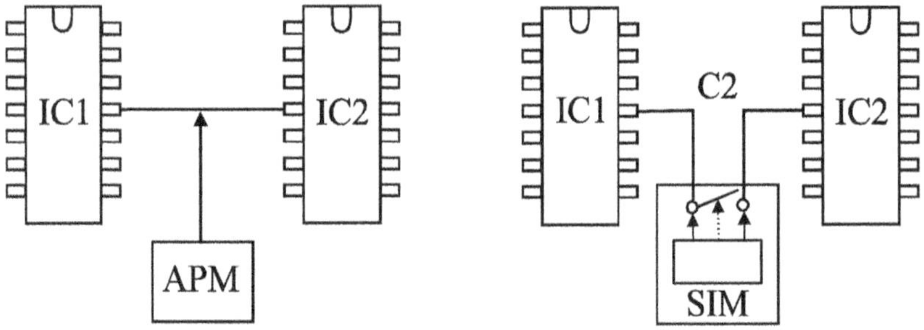

Figure 2.1-1. (APM) Active Probes module, (SIM) Socket Insertion module

Among the pin-level fault injection tools it is worth to emphasise the following:

- The MESSALINE injector from LAAS [ARLA_90], Toulouse (France), that implements both active probe and socket insertion techniques.
- The RIFLE injector [MADE_94], from the University of Coimbra (Portugal), that uses only the socket insertion technique.
- The AFIT injector (Advanced Fault Injection Tool) from the Polytechnic University of Valencia (Spain) [MART_99], which implements the active probe technique.

3. THE PIN LEVEL FI FARM MODEL

Following the FARM attributes, we briefly describe the most important concepts of the pin-level fault injection technique.

3.1 Fault model set

Fault models describe the fault using parameters such as the location and duration of the fault, as well as the type of perturbation (bridge, stuck-at, open line, etc.)

The physical perspective of this technique restricts the location of faults to the input and output target signals. However, the understanding of the fault model changes at upper levels. For example, at message level it is better to describe the fault effect (for example, a message corruption or "babbling idiot" incidence).

Fault timing parameters include fault duration and fault persistence: permanent, intermittent or transient. Short transient faults are frequently used, but the duration of these faults depends on the tool features.

3.2 Activation set

Depending on the objective of the fault injection campaign, it is a designer decision to trigger the injection randomly – without synchronism between tool and target, or with some type of synchronism. Although pin-level fault injection tools can inject faults triggered only by a random timing condition some type of synchronism between tool and target is also attractive. For example, when dealing with a microcontroller it will be interesting to trigger the injection if a predefined output of a peripheral is detected.

3.3 Readouts Set

To define how to read the system response is as necessary as to define the activation of the fault. In fact, if no suitable readouts are logged, the campaign will be totally rejected.

We can distinguish between readouts of the injection itself, and readouts describing the target behaviour in the presence of the fault.

The former group includes the effectiveness of the fault that at physical level is easy to understand. If we force a stuck-at-1 in a line whose logic level is one, the fault is ineffective.

The target behaviour is not always very easy to analyse. It is possible to analyse it by comparing the target outputs with a golden run execution. Other possibility is by means of a software monitor. Current VLSI designs include more and more elements inside the chip and reduce the number of output lines. Besides, a software monitor is cheaper and faster to implement, but it requires the target be able to maintain connection in the presence of the fault.

3.4 Measures set

Measurement is an off-line process carried out in function of the objective of the fault injection campaign. System coverage and Error Detection Mechanisms (EDM) coverages are typical measures. It is also important the validation of the target system that guarantees its correct specified behavior in a failure scenario.

4. DESCRIPTION OF THE FAULT INJECTION TOOL

4.1 AFIT – Advanced Fault Injection Tool

AFIT [GILB_97] was designed as a modular tool (Figure 2.1-2), with a personal computer handling the interface between user and injector. This software automates the injection process and enables a higher number of experiments to be carried out without supervision. AFIT injects faults at a frequency of 40 MHz, which is a considerable improvement over other fault injectors.

The tool is divided into five independent modules: Synchronization and Triggering module, Timing module, Fault Tolerant System (FTS) Activation module, Event Reading module and High Speed Forcing Injectors module.

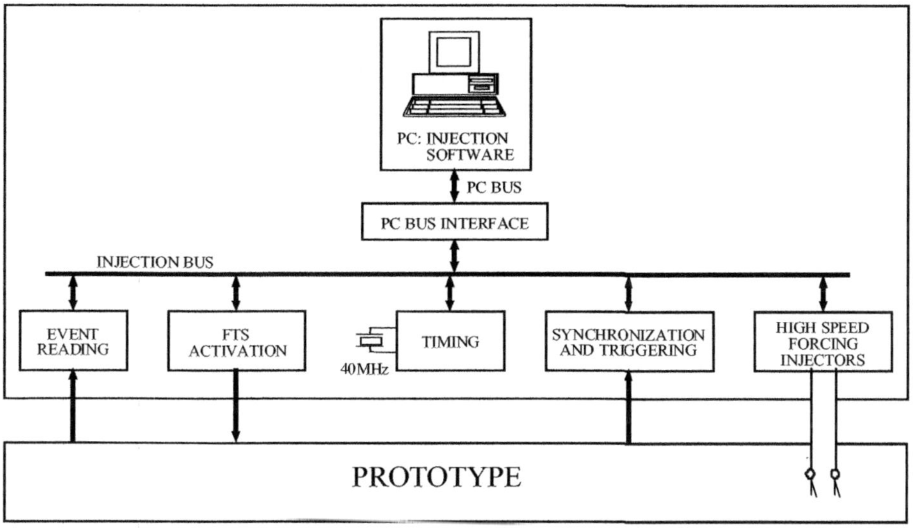

Figure 2.1-2 AFIT block diagram

Synchronization and Triggering: This module determines the beginning of the injection experiment. It is composed of two internal blocks, the Triggering Word Comparator and the Programmable Delay Generator (Figure 2.1-3a). After programming a Triggering Word (*TW*), and enabling the injection with the signal *IE*, this module samples some target signals (address, data or control signals) in order to determinate its Internal State (*IS*). When *IS* matches the programmed *TW*, and after a programmed delay, the second block of this module generates the Injection Triggering signal (*IT*), that activates the Timing module, which controls the fault injection. The programmed delay can be zero.

Timing: This module (see Figure 2.1-3b) generates the clock frequencies used by the tool. These frequencies are programmable up to a maximum, currently of 40 MHz. Furthermore it provides the Activation of the Injection signals (AI) that activate the High Speed Forcing Injectors.

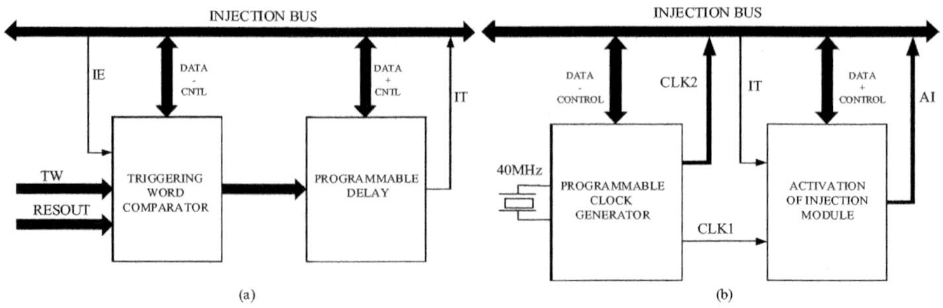

Figure 2.1-3. Block diagram of the Synchronization and Triggering module (a), and Timing module (b)

FTS Activation: The main function of this module is to prepare the target for each injection experiment, initializing it and activating its inputs with a predetermined Activation Vector AV. This module consists of two blocks (Figure 2.1-4a). The Programmable Reset Generator activates several Reset Input (RESIN) signals of the target. The active level of each signal is programmable. These signals are used to initialize the target. The other module called Activation Module programs several other activation signals (OAS) of the target in order to set them to a predefined initial state.

Event Reading: This module determines the response of the target system. This is done measuring the elapsed time from the activation of IT to the final recovery with an internal counter and a trace memory based on FIFOs (Figure 2.1-4b). Variations of the ERRV (Error Vector) signal, which characterise the response of the target after the fault, are stored in the State FIFO, and the value of the 24 bit counter, which indicates the time elapsed since the activation of IT, is stored in the Timing FIFO. The blocks Previous State Latch and New State Detector activate the Write enable of the FIFOs if any change in ERRV has occurred. If the counter overflows, the value of ERRV and the time associated are also written in the FIFOs. In this way, both fast and slow events in the ERRV signals can be managed due to this module acts as a transitional logic analyser.

The observation of ERRV signals requires special input probes to connect this module with the target system because the frequency of these signals can be up to 40MHz. These probes are based in a frequency compensated voltage divider that divides by 10 the observed signal, reducing by 10 the parasitic capacitance induced to each ERRV signal. This solution is very common in the design of input probes for oscilloscopes and logic analyzers.

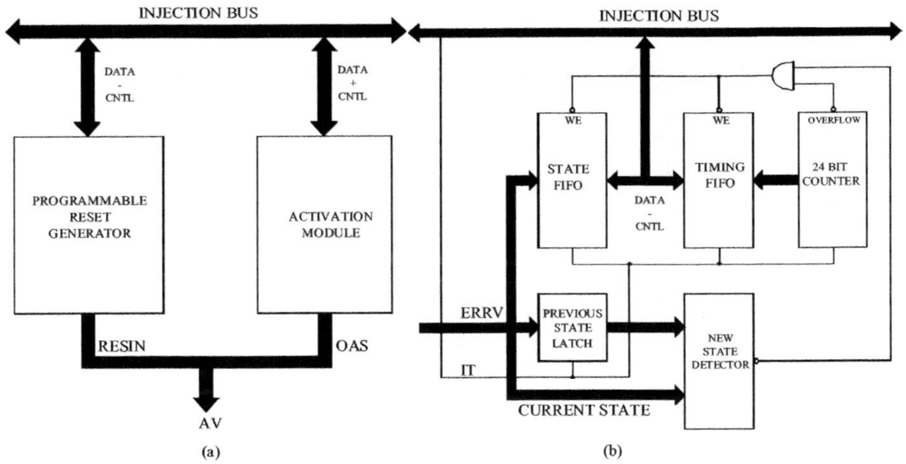

Figure 2.1-4. Block diagram of the Activation module (a) and Event Reading module (b)

High Speed Forcing Injectors: They physically force the faults. When the injectors receive the AI signals, the Injector Activation Logic (Figure 2.1-5) sets one of the gates of the transistors – the one on the top injects stuck-at-1 faults and the one on the bottom injects stuck-at-0 faults. The transistor selected injects the fault through the lines IOUT (Injection Outputs).

The other internal block, called Effective Error Detector, sets the MEE (Memory of Effective Error) only if the IOUT signal has forced the selected pin. In other case, the fault has been ineffective, and this signal is not set. This means that the experiment is not successful. The MEE signal can be read by the PC and is also one of the ERRV signals.

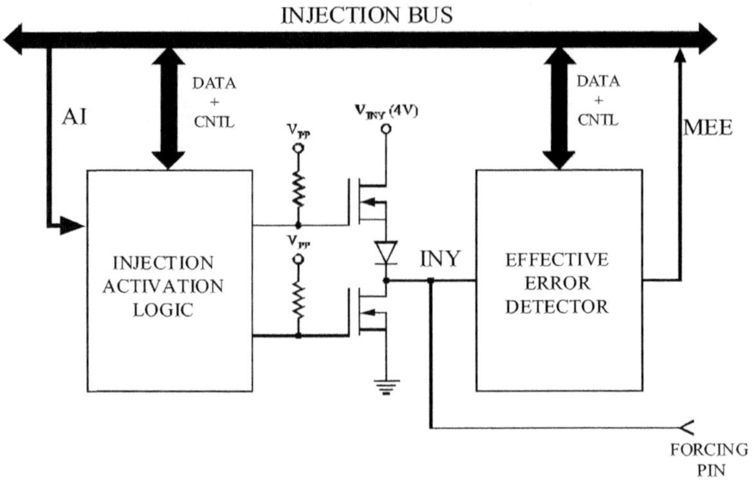

Figure 2.1-5. Block diagram of the High Speed Forcing Injectors module

Last version of AFIT was built in 1999 and only the Timing module and the High Speed Forcing Injectors were integrated in a compact tool with four independent probes. There are two static probes available (the fault location is decided before starting the injection campaign) and two more on-line assigned probes (the fault location is dynamically selected between 24 different pins). The tool starts the injection triggered by an external signal.

Currently, the Synchronization and Triggering module, as well as the Event Reading module are replaced with a new design – a tool adapter (Figure 2.1-6), suitable for triggering the injection and for implementing a software monitor of events for distributed real-time systems. The injection can be triggered synchronising tool and target or without synchronism using a random delay from the start of the experiment (usually the target reset).

The necessity of a software monitor for distributed real-time system was shown when using the tool in the validation of communication protocols oriented to safety critical applications. Information about the behavior of all units connected to the network or communication bus should be logged for a better understanding of the fault effect.

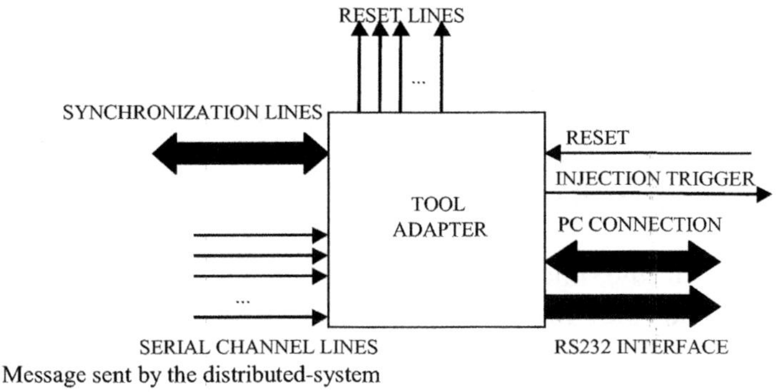

Figure 2.1-6. Tool adapter

4.2 The injection process: A case study

This section includes a practical case study on a real system [MART_99]. The real system and the methodology followed during the validation are briefly described.

4.2.1 System Description

The system selected was FASST [FASS_90], a fault-tolerant multiprocessor system composed by several fail-silent processor modules. The FASST prototype is made of two DPUs (Data Processing Units), a stable memory module and a channel for I/O computations. Each module is a Futurebus+ Profile B compliant device [SP109_94].

When a DPU fails, in case of error detection, the processor is immediately halted and the module is disconnected in order to prevent it from corrupting the rest of the system. Eventually the other modules will detect the module failure and will start the reconfiguration and recovery tasks, allowing the graceful degradation of the system.

The key to achieve fault tolerance are the processor modules, which include complex levels of fault tolerance mechanisms to ensure low error detection latencies and high error detection coverage. Any evaluation of FASST dependability must recognize that it strongly depends on the existence of fail-silent processor modules and on rollback recovery capabilities of the operative modules through the use of a stable storage system. FASST validation was made using pin level fault injection. In this system, since every processor signal is compared against its pair at every clock cycle, a minimal skew in these signals will cause comparison errors. It also incorporates a complex hierarchy of fault detection mechanisms where high-speed injection rates and low-resolution time in measurement instruments are required to evaluate the behavior of transient faults.

4.2.2 The injection campaign

The goals of the injection campaign are the estimation of the error detection coverages of the comparators, the parity and the watchdog timer included in the DPU and the calculation of the system recovery coverage. Another goal is to estimate the latency of the comparators in the detection of the error and the latency in the reconfiguration of the system. Furthermore, the time the module takes to halt once the error has been detected (in order to analyze the fail-silent characteristics) will be obtained.

Table 2.1-1 depicts the characteristics and the number of stuck-at injected faults according to their duration and the location. Permanent faults are active during 1.2 seconds. Transient faults are active during a random period between 100ns and 4μs. Intermittent faults are active periodically between 100ns and 2 μs in intervals between 1 and 65 ms. The injection point is randomly selected between the address, data and control busses, the host bus and the CSR bus.

Table 2.1-1. Number, location and kind of injected faults

		Injection point			
		Master processor	Slave processor	Host Bus	CSR Bus
Kind	Permanent	2640	2640	750	630
Of	Transient	2640	2640	750	630
fault	Intermittent	2640	2640	750	630
Total:		7920	7920	2250	1890
Total injected faults:					19980

Figure 2.1-7 shows the system block diagram and its more relevant signals and Figure 2.1-8 (and Tables 2.1-2, 2.1-3) shows all the possible reactions in the evolution of a single experiment.

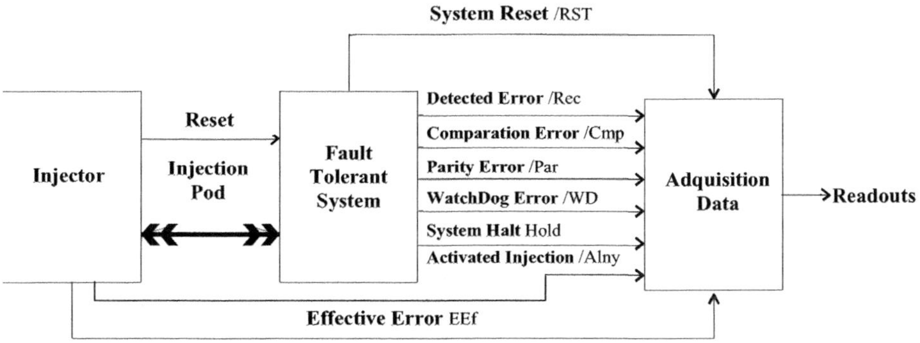

Figure 2.1-7. System block diagram

Readouts include the absolute timestamp of any event whenever a transition of the state machine takes place. The data selected to be stored are the reset signal from the Futurebus+; the injection activated and effective error signals from the injection tool; and the comparison error, parity error, watch-dog error and system recovered signals from the processor modules.

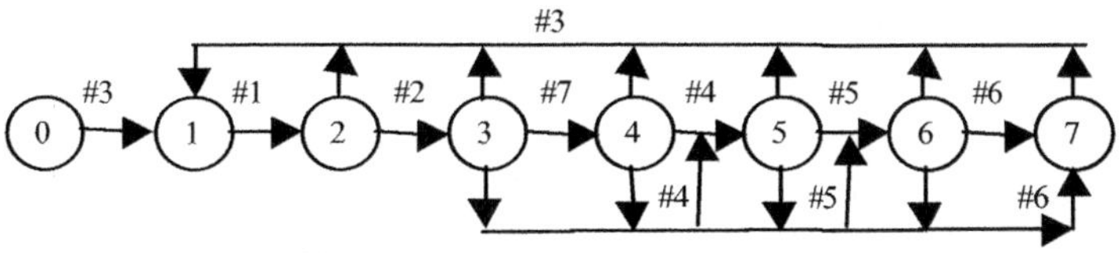

Figure 2.1-8. State machine and transitions

Table 2.1-2. Transition triggering words

	Transition triggering words							
	EEf	/Alny	/RST	/Par	/WD	Hold	/Cmp	/Rec
#1	X	X	1	X	X	X	X	X
#2	X	0	X	X	X	X	X	X
#3	X	X	0	X	X	X	X	X
#4	X	X	X	X	X	X	0	X
#5	X	X	X	X	X	1	X	X
#6	X	X	X	X	X	X	X	0
#7	1	X	X	X	X	X	X	X

Table 2.1-3. State machine that controls the readouts

Nº	State	Actions
0	Initialisation	Wait for Reset activation
1	Reset	Activation time acquisition
2	Activated injection	Starting time acquisition
3	Effective Error	Error effective latency calculation

4	Error detected by comparison	Comparator e. detection latency calculation
5	System halted	Global error latency calculation. Acquisition of /Par and /WD
6	Error detected by the other DPU	Recovery system latency calculation
7	Experiment ended	Wait for Reset activation

When the injection takes place (assertion of signal *Activated Injection*) the state machine evolves to a new state depending on the trigger that has been detected. According to Table 2.1-2, the detection of an error by the comparators, a system halt, a system reconfiguration or simply an assertion of the reset signal, which indicates that the experiment has ended and that a new one will proceed in a short time, can be signaled. The reset signal ensures that the automaton will always return to the initial state at the beginning of the next experiment, independently of the behavior of the previous experiment.

Several actions have been taken into account in the injection campaign:

- Suppress the redundant information (captures with the same signal states).
- Eliminate the glitches (in our case, we discard the readouts with glitches shorter that 90 nanoseconds).
- Enumerate the experiments using the reset signal.
- Discard the non-effective errors (we discarded 9% of the injected faults).
- Assess the effect of the failure over the system.
- Calculate the latencies.
- Format the readouts to be included in a spreadsheet to facilitate the statistical study of the results.
- Classify the experiments according to the cases we want to observe in the evaluation of the system. This step is quite important since, again, it allows us to discard the invalid experiments of the campaign. Table 2.1-4 shows the set of possible behaviors of the fault tolerant mechanisms included in the system. The last column shows the frequency of the event.

In our experiments we eliminated the cases that do not match with any of the events in Table 2.1-4. The use of a methodology has allowed us to detect a great deal of invalid cases due to possible noise induced by the injector or to bad probe connections.

Table 2.1-4. Fault Tolerant Mechanisms possible evolution

Cod	Event	%
1	An effective error is not propagated to the system. There is no failure	19.6
2	Non-detected error (a detection coverage failure)	0.1
3	System failure. The error is detected but the recovery mechanisms fail.	1.9
4	Comparison error detected	37.1

5	Parity error detected	28
6	Watch-dog error detected	2.7
7	The module is halted, however we can not guess the cause of the problem. The degradation process does not fail	4.8
8	Transient fault detected and recovered	2.4
9	System degradation due to a time overflow in the transient fault recovery process	0
10	The module is halted, however we can not guess the cause of the problem. The degradation process also fails	3.4

Making a straightforward analysis according to the error detection and system reconfiguration capabilities, we can build the dependability predicate graph of the system. Figure 2.1-9 shows the asymptotic error detection and system reconfiguration coverage of the system.

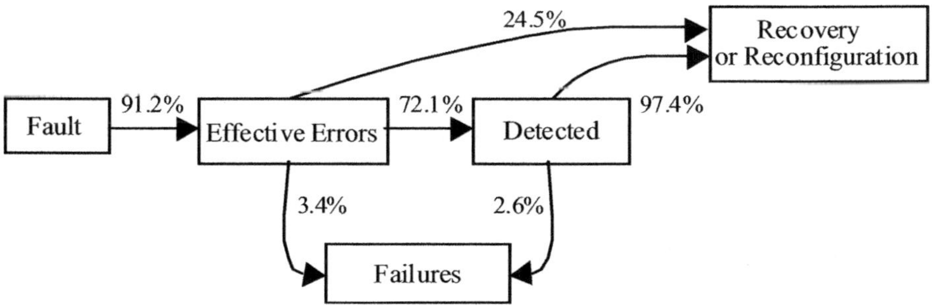

Figure 2.1-9. Fault-tolerance mechanisms predicate graph

One of the more important steps in the experimental validation is the measurement of the latency times of the error detection, module halt and system reconfiguration. Table 2.1-5 shows an example. This yields direct observation of the fail-silent characteristics of the module, and also the real-time capabilities of a simple application.

Table 2.1-5. Median of the latency times in nanoseconds

	Comparator	Parity	Watch-dog	Global
Error Detection	528	5242	1145	810
Module halt	6697.5	5242	1145	5170
System recovery	6,27E+08			

4.2.3 Execution time and overhead

The injection process can be divided into three parts: synchronization, trigger and observation time.

Synchronization: It is a very short time that depends on the delay from the beginning of the experiment to the event that triggers the injection (time, condition or both).

Trigger: The tool delay from the external trigger to the effective injection is short (25ns).

Observation time: Because of starting each experiment means restart the target free of faults, after forcing a fault there is no fear of leaving the execution running as long as it should be necessary (despite a system crash). This time is a key testing decision.

Usually, because pin-level fault injection works with the real system or prototype, the request period of time for the three parts is very low (one or two seconds is enough)

Other interesting point in fault injectors is the overhead question. An external tool does not produce overhead on the execution time of the target system. This type of tools has advantages and disadvantages. The greatest advantage is the lack of intrusiveness on the workload. But, on the other hand, it should be necessary overcome another type of intrusiveness exclusive of these tools. Disturbances slipped by the probes, especially in systems with high clock frequencies, can affect the correct operation of the prototype. There is a low probability that a component would be damaged in the process of fault injection using the active probe technique and it is almost impossible when using socket insertion.

5. CRITICAL ANALYSIS

First pin-level fault injection tools dates from twenty years ago and they have been used in many research works.

Injection damages on the target system (one problem of the pin-level fault injection tools) have been considerably reduced with new improved designs that take also into account the necessity of multiple injections, automation and synchronism with the target system.

As there is not overhead on the system under test and because physical faults injected in the real system or a prototype are closer to real faults, we find out that this technique attains important contributions to the validation and verification of fault-tolerant systems.

About the reusability question, this generic tool can be used with a wide range of prototypes. However, the problem of pin access must be resolved individually. That means, there are systems designed to be tested easily with all inputs and outputs signal well accessed. In other cases, we solve the problem using a test-clip adapter or intermediate resources.

To design and build a new tool is not cheap. The development time is long and a skilful hand is required. Moreover, voltages and frequencies of new VLSI designs reduce our control over the target system.

Although we solve the difficulties, and thinking on the advantages, it will remain the reachability question. Injections are always restricted to input and output signals, and the effect of the fault inside the chip at RTL level or combinational logic cannot be analyzed. Weakness parts of the VLSI are identified with statistic methods because there is not guarantee of a deterministic injection in a precise location inside the chip.

To trace the code execution in prototypes with internal memory is difficult and a detailed analysis of the built-in-self error detection mechanisms is not possible.

The outlook of the pin-level fault injection technique is not promising but it is still being useful. The reason is because there are aspects that remain without answer in spite of other fashioner techniques. Among the advantages of pin-level fault injection we find out the capability of reaching system parts inaccessible by software and the direct access to the input-output system including peripherals and communication busses. Besides, it is important to remark the lack of execution overhead as well as a fast injection process that allow us to inject a high number of faults automatically and in a short period of time. The real effect of physical faults inside the VLSI and their fault propagation is an interesting point to be carefully analysed.

Chapter 2.2

DEVELOPMENT OF A HYBRID FAULT INJECTION ENVIRONMENT

The Birth and Growth of "LIVE"

Leonardo Impagliazzo and Fabiomassimo Poli

Ansaldo Segnalamento Ferroviario, via Nuova delle Brecce 260 ,I-80147 Napoli- Itlay

Abstract: The reasons behind the choice of Ansaldo Segnalamento Ferroviario to design and develop a new tool for the experimental verification and validation of fault-tolerant, real-time, distributed digital systems is presented. The main features of LIVE (Low-Intrusion Validation Environment) and an example application are given.

Key words: fault models, fault injection, safety critical systems.

1. DEPENDABILITY TESTING AND EVALUATION OF RAILWAY CONTROL SYSTEMS

Modern railway and metro lines control systems [MONG_93], [HACH_93], [HENN_93] perform signaling and automation functions. The transition from relays to computer-based systems, due to an increased demand in reliability and performance, led to stress the need for the design and assessment of the safety of completely new architectures[1]. The CENELEC norms [EN50126] [ENLE_A], [ENLE_B], approved as European standards, require:

1. The use of verification and validation processes in all phases of the life-cycle of the system;

[1] While these systems may deeply differ in the fault tolerance mechanisms adopted, in general they are based on redundant distributed architectures that make use of microprocessors boards [11].

A. Benso and P. Prinetto (eds.), Fault Injection Techniques and Tools for Embedded Systems Reliability Evaluation, 41-48.

2. The demonstration of compliance with quantitative safety targets that depend on the criticality of the system functions.

These recommendations have brought about the definition and development of new methodologies for safety design and assessment.

In the validation methodology, two phases are of primary importance:

- Dependability testing *(fault removal)*, aimed at the discovery and correction of design errors in the management of the faults in hardware-software system prototypes;
- Dependability evaluation *(fault forecasting)*, combining the experimental results with analytical models in order to evaluate the HFR (Hazardous Failure Rate) [ELEC_B], [AMEN_96_A], that is to be kept under specified levels, depending on the criticality of the function performed by the system[2].

The CENELEC norms do not specify what techniques must be used, but the methodology, the tools, the results and more generally the entire design and verification and validation process have to be approved by an independent safety Assessor [EN50126].

The next section highlights the "requirements" that leaded us to choose a hybrid fault injection environment for the validation of railway signaling systems [AMEN_96_D].

2. BIRTH OF A VALIDATION ENVIRONMENT

In the system life cycle of a real time and fault tolerant system a critical step, for both design and verification, is the prototype phase [IYER_95]: i.e., at the end of SW and HW design phases the system SW and HW become available and must be integrated together. In this phase the ability of the system to behave as expected and the correct implementation and effectiveness of Error Detection and Recovery Mechanism (EDRM) have to be demonstrated. Of course only an accurate design in the previous phases can ensure the required system dependability, but the possibility to actually demonstrate real time and fault tolerant features before the prototype phase can be argued [AMEN_96_D]. So the first requirement (or constraint) for our validation environment is:

R1) A HW/SW prototype of the system is available.

[2] E.g., for system elements which have the highest level of Safety Integrity (SIL4), the CENELEC norm EN 50129 - in its "prestandard" versions - prescribed a value of the dangerous failure rate per hour and per element less than 10^{-10} (less than 1 failure in 1 million years!) in Continuous/High demand mode of operation.

In order to be able to functionally test an equipment, even in absence of the other systems with which it will be integrated, environment simulators are normally developed together with the system prototype. The environment simulator shall be able to create meaningful operative conditions for the system; it shall allow to set-up and record workload scenarios (sequences of inputs), to repeat a defined scenario, and to record the system behavior (sequence of outputs). Our validation environment requires that:

R2) An environment simulator is available for the system.

A standard feature of embedded and commercial HW boards is to provide a way to upload the SW. In some case the loader is part of a more powerful "on target" monitor debugger that allows memory dumping and breakpointing. In the case of industrial systems, it is also available the possibility to dump through a serial port the history of events and errors recorded by the system during operation, and sometimes the logical state of the system. The third requirement for our validation environment is:

R3) The system under test allows to upload its SW and to download historical data and state.

It is worth noting that the above requirements (*R1*, *R2* and *R3*), which can be a serious constraint for a research project, do not add any cost to a standard industrial project, as they are consolidated part of the design process.

As one of the main objectives of the prototype phase is to evaluate performances and EDRM, suitable tools have to be identified. Table 2.2-1 below summarizes the results of a comparison among common monitoring tools [AMEN_96_D]; "dark gray" cells identify fully implemented features, "light gray" cells identify partially implemented features.

As shown in the table, SW debug monitor and Logic Analyzer define an optimal combination if properly used. Herein a few aspects are underlined for the Logic Analyzer:

- The ability to trigger on well-defined HW conditions (bus states or sequences);
- The ability to measure time with very high resolution (nanoseconds);
- The negligible intrusion.

The ability to trigger on well-defined HW conditions allows catching and tracing any event (e.g., the activation of an error detection SW function), as well as complex bus sequences. Moreover Logic Analyzers can provide external triggering signals that can be used to activate external HW when the defined trigger condition is satisfied: this feature plays an essential role for fault injection (see section 3.1).

The ability to measure time with very high resolution is essential to evaluate performances (e.g., measuring the execution time of a set of system diagnostics in different workload scenarios). In a fault injection experiment, it is possible to measure the fault dormancy, the error latency, the negation time, etc. [AMEN_96_C] (e.g., measuring the time intervals between the corruption of an incoming message, the detection of the error by the CRC check function, the sending of the Not-Acknowledge message, and the receiving of the repeated correct message from the sender).

The negligible intrusion is ensured by the high electrical impedance of the Logic Analyzer probe connected with the microprocessor pins. This feature is one of the most important in the validation of safety critical system: it allows proving without doubts that the system under test is not affected by the testing environment, and therefore the measures and results are not biased.

Table 2.2-1. Comparison among monitoring tools

	In-Circuit Emulator	**ROM Emulator**	**SW Debug Monitor**	**Logic Analyzer**
Download Code				
Start at 'Address'				
Single Step				
Memory Fill/Dump				
Memory Disassembling				
Examine/Change Memory				
Examine/Change Registers				
Real-time Trace				
Software Breakpoint				
Hardware Breakpoint				
Multi-level Breakpoint				
Reset Prototype				
Monitor General Purpose Logic				
Control General Purpose Logic				
Source Debug Access				
Timing Analysis				
Intrusion				
Multiprocessor Analysis				

What above lead to an obvious choice:

R4) The validation environment shall relay on logic analyzer(s) for monitoring and synchronization.

The major limits in using a Logic Analyzer are the high cost of such measurement equipment, the complexity to insert it in an automated fault injection environment, and the low level monitoring. The low level monitoring, which can be very helpful in some case, limits the time interval in which the system can be observed (generally no more than 1 million sample can be acquired). The high cost is compensated by the availability from the supplier of up-to-date SW supports for almost all microprocessors; this simplifies the porting of the validation environment to new systems.

The most effective approach to verify the correct implementation of EDRM is to inject fault and monitor the system. In order to do this, an appropriate (representative) fault model has to be chosen. Our simple but reasonable assumption is: "any fault within the system can be represented as a corruption of the contents of a register or of the data moving among registers; the registers are those defined in the programming model of the system".

With this simple assumption in mind, it is possible to define a suitable fault injection technique. It is out of the scope of this chapter to compare fault injection techniques [Chapter 1.1]; only the reasoning behind our solution is presented. The analysis of the available fault injection techniques leaded to the conclusion that a single fault injection technique cannot be appropriate in all cases. The following requirement was added to our validation environment:

R5) A combination of both hardware and software implemented fault injection shall be used.

Hardware implemented fault injection can corrupt data transferred on the system bus with negligible intrusion. In many cases, it can emulate corruption of internal register by corrupting the data transfer. As the time synchronization is at bus level, it is very accurate. Unfortunately special purpose hardware is required and the technique can fail in emulating some device internal faults, particularly if permanent (e.g., corrupt permanently a bit in a CPU data register).

Software implemented fault injection can easily corrupt any location visible in the programming model (e.g., internal registers in the programming model of a microprocessor, a serial communication controller, any location in RAM, etc.). The simplest and less intrusive way is to activate a non-maskable interrupt service routine containing the corrupting code. It is worth noting that the high intrusion of certain software implemented fault injection environment is mainly due to the SW monitoring not the SW injection; the Logic Analyzer solves this limitation, as well as the problem of the weak time synchronization.

The Fault Injection Board (FIB), developed in a joint project with the "Politecnico di Torino", supports our hybrid fault injection technique [AMEN_96_C].

The above-identified requirements cannot fully characterize an industrial validation environment unless all the elements are combined together providing a unique well-documented interface. As a consequence of this:

R6) The monitoring and fault-injection features shall be integrated in a fully automated environment

The LIVE (Low Intrusion Validation Environment) toolset, developed between 1995 and 1997, satisfies the identified requirements. It was presented in [AMEN_96_C] where an example application was also given[3].

From 1996 to 2000 the LIVE environment was applied to the validation of several railway control systems based on two out of two and two out of three architectures. It was successfully ported on several Motorola processors (MC68HC11, MC68302, MC68360, MC68040, etc.). Rather than fault forecasting, the main objective was the testing of EDRMs.

3. THE EVOLUTION OF "LIVE"

In 2000, within the framework of a join project between Union Switch & Signal and Ansaldo Segnalamento Ferroviario, LIVE was chosen as the fault injection environment to measure the fault coverage of a fault tolerant system based on single self checking architecture (see section 4). University of Virginia was the research partner, in charge of defining the overall approach and the fault model.

A gate level VHDL model of a RISC processor was used to define a generic fault model applicable also to CISC microprocessors. The LIVE approach to corrupt register contents and data transfers was improved as a result of this analysis and of detailed investigation on memory mapped devices (clock, serial communication controller, bus controller, digital I/O, etc.). The following additional corruption models were identified:

a) Terminate fetch and execute – the effect of fault is to stop the CPU, so that no new instruction is fetched or executed;
b) Multiple corruption – the effect of fault is to corrupt at the same time several CPU registers;
c) Instruction Register corruption – the effect of fault is to corrupt a subset of instructions;
d) Wrong interrupts – the effect of fault is the generation of a generic interrupt (any vector), from a device which is not programmed to do this;

[3] In particular, the impact of the workload on fault coverage and latency measures was investigated: the interesting result was that, while fault coverage and fault dormancy strongly depend on the workload, the error latency is mainly related to the EDRM. Moreover, a combination of appropriate simple scenarios can produce more reliable measures than a very complex workload.

e) Unexpected and missing interrupts – the effect of fault is a spurious or missing generation of the programmed interrupt, from a device.

All these faults can be either transient or permanent and can happen at any time in the workload scenario. The LIVE environment was potentially able to implement the above additional fault models, with the exception of the instruction register faults: new hardware was added to the FIB to be able to recognize the instruction fetched on the data bus and to corrupt it before the CPU completes its bus cycle.

The fault model is characterized by several properties: the corruption model (how many bits in a word must be "flipped" or "stuck-at"); the duration (80% transient and 20% permanent); the time of injection (uniform distribution in the scenario with a 100ns resolution). Once defined the characteristics of the faults that can affect each device in the system under test, the distribution of the faults was based on the result of VHDL simulation and on reliability prediction methods.

An algorithm was defined to generate the list of faults to be injected. A configuration file allows adapting the fault parameters to any system. The LIVE environment was improved to be able to inject all faults in the fault list; some examples are presented in the following subsection to show the new features.

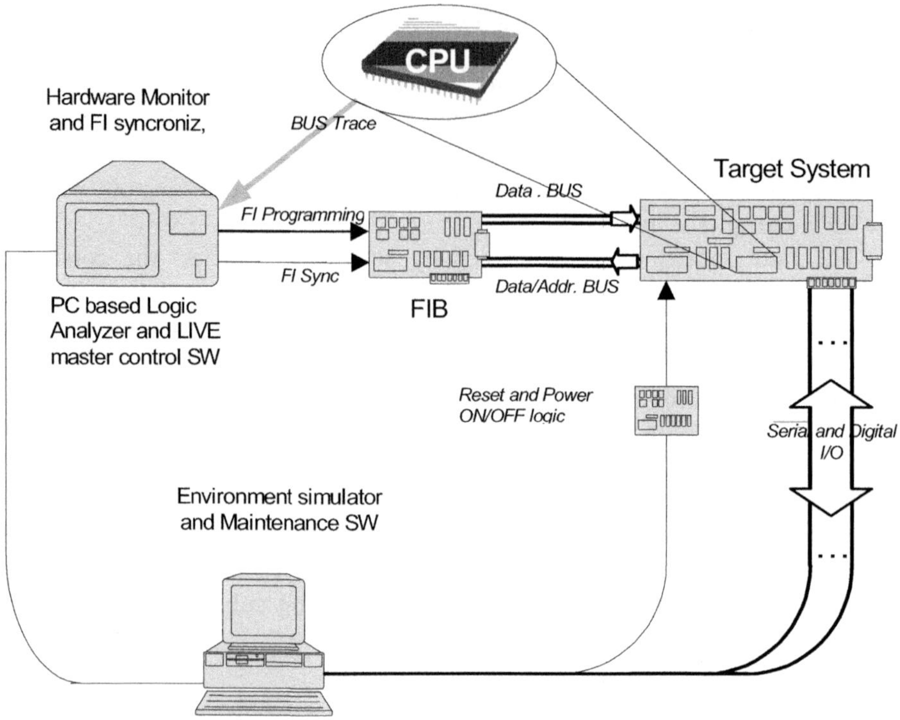

Figure 2.2-1. LIVE Structure

Figure 2.2-1 describes the structure of LIVE. Its main components are:

1. The prototype of the target system running the actual system SW;
2. The PC running the maintenance SW, the environment simulator and the logic to reset and power-ON/OFF the target system;
3. The PC based Logic Analyzer that includes the SW for FIB programming and the LIVE master SW to automate and control the whole fault injection process;
4. The enhanced FIB (Fault Injection Board) to inject bus faults and generate non-maskable interrupts.

3.1 Two examples of automation

LIVE is a complex set of tools; we chose two simple examples to give an idea of its features.

As an example of interrupt based injection consider a permanent fault within a CPU register: e.g., the Motorola MC68332 D0 data register. The fault is characterized by the following parameters where the second field is the time of injection in microseconds and the last is the random mask that identifies the bit(s) to corrupt:

```
334 92578987.5 Perm Reg.File(int) CPUregD0
00000000000000000000000000000010
```

The LIVE environment automatically generates both the script for the master control SW and the interrupt routine to be loaded on the target system; assuming a stuck at 0 is selected (50% probability), the routine code is only two instruction long with a few microseconds intrusion:

```
ANDI.L #0xFFFFFFFD,D0
RTE
```

The LIVE master control SW executes "scripts" that can contain thousands of experiments to be run in sequence. The script that lists the actions to perform in order to realize the experiment described above is what follows (comments are embraced by the sting "**"):

```
** Load the Logic Analyzer (LA) with the appropriate configuration          **
@NMODULE CPU32             ** Motorola CPU32 bus trace support SW **
@LCONFIG c:\FaultTst\systems\reg_D0_P.tla     ** Inject reg. D0 permanently **
**          ________________Experiment number 000334________________     **
** Set the time of injection from start-up in picoseconds                 **
@STRIGGER 2.1.1 92578987500000
** Connect to simulation PC. Enable env. simulation. Power on the system    **
** and load the interrupt code using the maintenance SW.                    **
** At the end, power off the system                    **
@EXE rsh SimPC Simulator.exe LoadInt.txt \codInt\Int000334.txt
** Program FIB to inject an interrupt on Logic Analizer trigger signals.    **
** Status of address bus is not relevant. Interrupt is not maskable.      **
@EXE ficonfig.exe /addr=FFFFFFFF /signal=1xxxxxxx /opctrl=start
         /optype=INTERRUPT /intlev=0
** Enable LA state machine: time of injection will be counted from power-up **
** Samples are acquired starting from the injection up to available memory  **
@START nopause
** Connect to simulation PC. Enable env. Simulation. Power on the system.   **
** and record its serial and digital I/Os with the environment simulator.  **
** Before end of scenario store the internal events and application state  **
** using the maintenance SW                          **
@EXE rsh SimPC Simulator.exe Workload.txt \simlog\smlg000334.log
        \eventlog\evlg000334.log \userlog\uslg000334.log
** Wait for the completion of the selected workload scenario               **
@WAIT 290000
** Stop Logic Analyzer state machine and FIB injection                   **
@STOP
@EXE ficonfig.exe /opctrl=stop
** Save Logic Analyzer acquired samples                       **
@SAVE C:\FaultTst\Listing\lalg000334.txt
```

The state machine of the Logic Analyzer is programmed to be able to inject again the fault (repeat the interrupt) after each write operation on register D0 (permanent fault); it generates the activation signal for the FIB that actually implement the interrupt bus cycle; the non-maskable interrupt vector points to the interrupt routine loaded in the system memory before the experiment starts. The instructions that can lead D0 to change are recognized by masks defined for Motorola 68000 family processor; as the available masks are limited, the interrupt can be generated more than needed. The fault is kept up to the end of scenario or the system stop/reset due to detection of the fault. Permanent faults on heavily used CPU registers are likely to be detected, so limiting the intrusion of the multiple interrupts. The Logic Analyzer acquired trace provides the evidence of the injection and allows detailed analysis of the effect of fault in the system; these features are essential to make the fault injection results auditable.

The environment simulator and the tool for maintenance provide the set of experiment readouts: both the external behavior and the system own view is available. LIVE provides tools to compare the experiment readouts (text files) against the result of the Gold-run execution (scenario running without faults): such tools are often complex and application dependent. The tools allow judging if there is a failure, or the fault is either detected, tolerated, or no-responsive. Moreover, it is possible to identify the most effective EDRMs, understand if the fault is detected by the expected EDRM or only by chance, and so on.

It is worth noting how the experiment number appears in the name of all files relevant to the experiment (interrupt routine, readout files, etc.); the importance of this "detail" appears clear when thousands of experiments are executed, gigabytes of files are recorded and you are looking for everything is relevant to experiment N° 084016.

As an example of injection on the bus, consider the permanent fault within the Arithmetic Logic Unit that corrupts all ADD operations from register D1 to any data register in SUBs, i.e.:

`ADD.L D1,Dx` corrupted to `SUB.L D1,Dx`

In our fault model this is mapped as Instruction Register fault; its parameters are:

```
521 58606372.0 Perm ALU(int) CPUregIR 1101XXX010000001 0100000000000000
```

The last two fields identify respectively the `ADD.L D1,Dx` instruction to corrupt and the *XOR* mask that "flips" bit 14 (`ADD` turns into `SUB`).

In this case no interrupt is needed and LIVE only generates the script file. The script file has the same structure shown in the previous example, the only differences are:

a) The Instruction Register state machine is used.

```
@LCONFIG c:\FaultTst\systems\reg_IR_P.tla    ** Inject Reg. IR permanently **
@STRIGGER 2.1.1 58606372000000
```

b) No interrupt code is loaded.
c) The FIB is programmed for data bus injection (as data bus recognition and corruption is required, enhanced FIB is used; 16 bits data bus is physically connected to FIB input address bus).

```
@EXE ficonfig /addr=FFFFDE01 /data=xxxxxxxxxxxxxxxxx0xxxxxxxxxxxxxx
        /signal=1xxxxxxx /opctrl=start
```

This kind of injection is able to create really complex faulty behaviors: e.g., the possibility to mask some bits in the recognition mask allows corrupting all opcodes with a certain addressing mode. The major limit is that extension words that match the recognition mask are also corrupted.

Starting from the fault list, the process to generate the script for the master control program is fully automated. The HW set-up has to be changed if moving from interrupt based to data bus injection, but grouping together experiments that share the same HW set-up, they can run without assistance 24 hours a day. As the fault script generation and the readout analysis is not performed within the fault injection environment, the time duration of each fault injection experiment is that of the selected workload scenario; only in case of interrupt injection, several additional seconds are needed to load the interrupt routine into the system.

4. EXAMPLE APPLICATION

The most extensive application of LIVE was in 2002, for the safety assessment according to CENELEC standards of the three computing platforms of the Metro Copenhagen Automatic Train Control System (ATC)[4]. Fault injection was selected as method to estimate the fault coverage: coverage of at least *0.999928* was required to meet the Hazardous Failure Rate (HFR) of the entire ATC. It was estimated that *32000* responsive experiments were needed to achieve the target. An environment simulator was built to emulate the entire area of the Ørestad station and the train movements at system interface (*16* vital digital outputs, *48* vital digital inputs, *32* non-vital outputs, *32* non-vital inputs, *3* serial links with vital protocol and *1* serial link with non vital protocol). The workload, which lasts around six minutes, includes:

a) System power-on and start-up;
b) Light workload (no trains moving or approaching the area);
c) Heavy workload (up to three trains moving in the area, stopping and leaving the station);
d) Light workload (no trains moving, system internal info download).

[4] The wayside interlocking platform (MICROLOK II) of the Ørestad station was selected as system under test. The MICROLOK II system (produced by US&S) is based on a single self-checking architecture and it is widely used around the world for railway safety critical applications.

Up to *6* LIVE platforms were used to perform more than *50000* fault injection experiments in three months. The fault coverage achieved was *0.99995*.

Table 2.2-2 below shows the results for permanent faults (20% of the total injected faults), providing for each fault class:

- The % of faults to be injected in each class in accordance with the actual fault rate.
- The % of *no-responsive* faults (i.e., faults which injection does not produce any effect); these experiments are discarded and replaced by new others in the same class.
- The % of *covered* faults (i.e., the system detects the fault, assigns a safe state to vital outputs and resets).
- The % of *tolerated* faults (i.e., the system might or might not detect the fault, but assures safer outputs without resetting).

The analysis of the complete results will be presented in future papers.

Table 2.2-2. Fault injection experiments results for permanent faults

Major Fault classes	**faults %**	**no-resp.**	**covered**	**tolerated**
CPU (MC68332) - Instruction Register faults	0.67	24.66	73.48	1.69
CPU (MC68332) - All other registers faults	0.55	1.49	98.30	0.21
Other CPU (MC68332) faults	0.30	0.00	100.00	0.00
MC68332 Internal Memory Mapped Devices Register Corruptions	0.90	25.25	74.31	0.43
MC68332 Internal Memory Mapped Devices Interrupts	0.24	21.13	66.90	11.97
Memory Mapped External Devices - Bus	7.96	19.60	74.54	5.85
Memory Mapped External Devices - Interrupts	2.65	8.77	87.90	3.32
RAM and ROM Memories	6.74	0.00	99.92	0.08
Totals for permanent faults	*20.00*	*12.96*	*83.66*	*3.37*

5. CONCLUSIONS

This chapter presented how a hybrid fault injection environment can be used to experimentally validate the capability of a complex railway signaling system to meet its quantitative safety target. The ability of LIVE - the validation environment of Ansaldo Segnalamento Ferroviario - to perform such a task was proven, and the results of its use in the framework of an actual industrial application were summarized.

Chapter 2.3

HEAVY ION INDUCED SEE IN SRAM BASED FPGAS

Marco Ceschia[1,2], Marco Bellato[2] and Alessandro Paccagnella[1,2]

[1]*DEI, Università di Padova, via Gradenigo 6/a, I-35131 Padova-Italy*
[2]*INFN, Sezione di Padova, via Marzolo 8, 35131 Padova, Italy*

Abstract: In this chapter we present an experimental methodology to study the Single Event Effects (SEE) in Field Programmable Gate Arrays (FPGAs) by using heavy ion beam irradiation. We present also some results obtained at a Van Der Graaff Tandem accelerator facility. Test methodology is based on the implementation of four Shift-Registers (SRs), two of them by using Triple-Modular-Redundant (TMR) technique. The SRs functionality is continuously checked during irradiation and errors are immediately and automatically logged and timestamped by the control system. Irradiation experiments have been performed on commercial FPGAs manufactured by Altera. These devices are based on static RAM configuration memory. No Single Event Upset (SEU) was detected in the shift registers upon irradiation. Instead, the predominant effect of heavy ion irradiation was the Single Event Functional Interrupt (SEFI). SEFIs are likely induced by SEUs in the configuration memory. No destructive latch-up has been observed. During irradiation, supply current has been observed to increase almost linearly with ion fluence, probably due to progressive SEU-induced driver contentions. The configuration memory cross section has been calculated.

Key words: FPGA, irradiation, COTS, SEU, SEFI

1. INTRODUCTION

The technological evolution of Commercial-Off-The-Shelf (COTS) Field Programmable Gate Arrays (FPGAs) is so rapid that these components are

A. Benso and P. Prinetto (eds.), Fault Injection Techniques and Tools for Embedded Systems Reliability Evaluation, 41-48.

progressively replacing ASIC devices, thanks also to the rapid evolution of the design techniques and Electronic Design Automation (EDA) tools. In general, the use of COTS in environments with high ionizing radiation content (such as in satellite space missions) in place of radiation-hardened counterparts is convenient because of the state-of-the art features of COTS, including lower cost, higher velocity, lower power consumption, higher density, etc. These advantages derive from the rapid evolution of the commercial CMOS technology, which is quickly approaching the 100nm minimum size limit. However, use of COTS components may be possible only once they are tested versus ionizing radiation damage. Hence, it is worth investigating the behavior of COTS FPGAs in radiation hostile environments to identify and quantify problems and limitations coming from radiation effects, and possibly relying on system level solutions for an enhancement of their reliability. Some previous works [WANG_99], [FULL_00], [OHLS_98], [KATZ_97], [KATZ_98] have addressed the impact of radiation damage on the performance of COTS FPGAs during the past years. Both total dose damage and sensitivity to Single Event Upset (SEU) of different FPGA families have been tested. However, to our knowledge, no attention has been paid to evaluating the radiation effects on Altera devices, which deserve high interest because of their internal architecture and their widespread use in non-hostile environments. In this work we have investigated heavy ion effects on Altera FLEX10K type components, which are commercial components without any radiation qualification.

2. EXPERIMENTAL SET UP

Our experiments were performed on SRAM based FPGAs. We have chosen the device type from the Altera FLEX family. The selected device was the EPF10K100 featuring roughly 100,000 equivalent gates and consisting of ~5,000 Logic Elements (LE), ~620 Logic Array Blocks (LABs), and ~25,000 embedded RAM bits.

The heavy ions used throughout our experiments are listed in Table 2.3-1. Device surface was kept orthogonal with respect to the ion beam axis, so the equivalent Linear Energy Transfer (LET) coefficient was exactly the ion LET. We would like to remind here that the LET corresponds to the ionizing energy lost per unit path length by the impinging ion, i.e., the amount of electron-hole pairs generated along the ion track. Prior to radiation testing, the package of each FPGA chip was submitted to a mechanical-chemical delidding process in order to expose the stripped die surface to the ion beam. All radiation tests were performed at the Tandem accelerator of the Laboratori Nazionali di Legnaro – Istituto Nazionale di Fisica Nucleare and Università di Padova, Italy.

In order to obtain small experimental errors, low ion fluxes were used wherever possible and the most significant results were obtained with a flux below 20 ions·$s^{-1}cm^{-2}$. The dosimetric system used at the beam line permitted the measurement of such low fluxes by employing a battery of reverse-biased silicon diode detectors, i.e., PNN structures with a sensitive area of 0.16 cm^2, with a pulse counting readout electronics. The diodes, distributed at the vertices of a 4cm square located immediately upstream of the target plane, surrounded the DUT. The ion flux measured by using the upstream diodes was periodically checked with a set of diodes packed into 1 cm^2 exchangeable with the DUT. The particle flux over the full area covered by the diodes could be made quite uniform by a magnetic beam defocusing system. With this dosimetric experimental setup we were able to measure ion fluxes in the range between 10 $cm^{-2}s^{-1}$ to a few 10^5 $cm^{-2}s^{-1}$. The ion fluence was actually read every 2 s.

The experimental setup has been developed in order to test FPGAs in a dynamic mode, that is, by exposing a Device Under Test (DUT) with a running circuit to an ion beam. This approach appears as a natural choice to emulate the radiation effects on operating components, even though the results are known to depend partially on the kind of circuit implemented in the DUT. Still, it should be complemented by a static mode of operation, that is, a configured and ready-to-run (clocked, without toggling inputs) DUT should be exposed to the ion beam. In this way we may collect important information on the single bit sensitivity to radiation. Remarkably, this second operation mode relies on the possibility of reading back the FPGA configuration memory, to check for SEU induced bit flips, and compute the SEU cross section of the whole device. Unfortunately, this feature is not available to the retail consumer and, consequently, the static test has not been performed.

Table 2.3-1 Radiation sources used in the experiments

Source	Energy [MeV]	LET [MeV·cm^2/mg]
^{12}C beam	88	1.6
^{16}O beam	109	2.8
^{19}F beam	117	3.8
^{28}Si beam	158	8.5
^{48}Ti beam	164	21
^{58}Ni beam	209	29
^{79}Br beam	194	40
^{127}I beam	257	62

In the dynamic test mode we implemented in the FPGA four Shift-Registers (SR), two of them designed with Triple Module Redundancy (TMR), operated at a clock frequency of 16 MHz. The SR outputs are compared two by two to check equivalence at each clock cycle and detect possible radiation induced errors. In particular, for Q1 and Q3 one TMR SR is compared with one non-TMR SR, while for Q2 two TMR SRs are compared.

Besides, a clock generator (TRIG) outputs a heartbeat reference signal at a frequency of ≅30 kHz, ideally suited for long distance transmission from the accelerator experimental hall to the shielded control room over inexpensive copper cables via RS422 transceivers. Three output signals (Q1, Q2, and Q3) replicate the heartbeat signal whenever the circuit is working properly. Schematic of this design is drawn in Figure 2.3-1 and TRIG, Q1, Q2, and Q3 signal shapes are drawn in Figure 2.3-2.

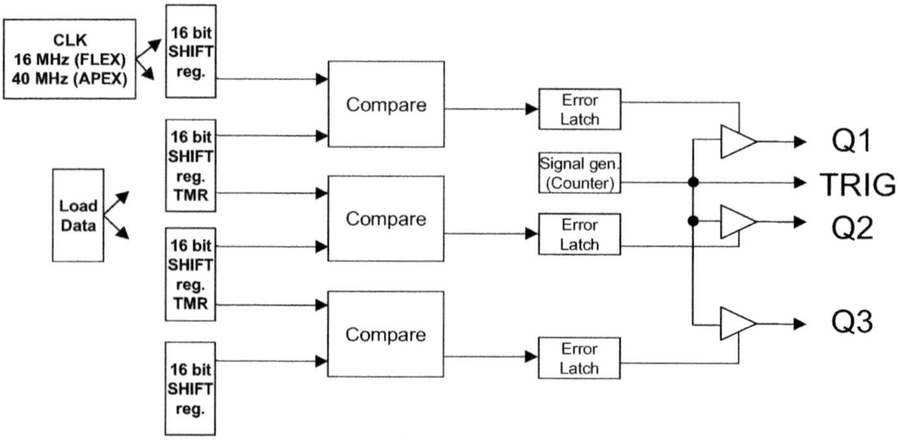

Figure 2.3-1. Design implemented in the FPGA under test.

When a Single Event Upset corrupts one SR (user circuitry), the corresponding output signal differs from the reference signal: the abnormal condition is immediately detected via a monitoring board controlled by a LabView program and the condition is time-stamped and logged. The SRs are then automatically re-loaded and the system is ready to detect another SEU at the next TRIG signal. Every cycle of the TRIG signal allows to detect one SEU, so the ion flux must be set low enough to avoid multiple SEUs during one TRIG period. Noticeably, the measurement system allows detecting two SEUs only when they occur in different SRs.

During irradiation the FPGA supply current (Icc) has been continuously monitored. Aiming to measure the Single Event Latch-up (SEL) cross section of the device, a second test procedure has been set up. This measurement has been carried out using an in-house designed power supply, stopping the supplied voltage when the current absorption exceeds a preset value (~800 mA in our case) and automatically re-applying it after ~300 ms from the interrupting event. Such a prompt inhibition of the power supply avoids a possible permanent damage or even a catastrophic degradation of the FPGA circuitry when a SEL happens. The preset value has been chosen at 800 mA, this value being about 3 times the current (~250 mA) absorbed by the FPGA under test during normal operation and below the maximum current allowed by the power dissipation capabilities of the device. We have based our SEL measurement procedure on the recording of the number of times that the supply current exceeded the upper limit of Icc=800 mA during a given irradiation time. This irradiation period has been determined during the experiments in order to have a minimum number of counts of about 20.

3. SEES IN FPGAS

3.1 SEU and SEFI

Usually, at the very beginning of an irradiation run, our LabView-based control system recorded a large number of errors. These errors were not caused by an equivalent number of SEUs but were induced by one or more Single Event Functional Interruptions (SEFIs); in fact the testing circuit, implemented in the DUT FPGA, could not reload the SR pattern automatically. We could observe successive SEFI induced errors by noting that the four FPGA output signals (see Figure 2.3-2) progressively disappeared during irradiation. Sometimes the missing signal suddenly re-appeared, but this could happen some seconds after the signal disappearance, that is, over a time much longer than the transient associated to a single incident ion. In this condition, no soft action was able to restart properly the circuit even if the ion beam was stopped. After a switching off/on cycle and reloading the entire configuration memory, the initial FPGA functionality could always be restored, indicating that no permanent damage was produced by the ion irradiation. We cannot state, of course, that SEUs in the flip-flops of the SRs never happen, just because SEFIs happened long before any single SEU could be observed. This result agrees with the measurements and results already available for Actel and Xilinx SRAM-based FPGAs [WANG_))], [FULL_00].

We have classified SEFI events as follows. When the first signal disappeared (Q1, Q2, or Q3) we took it as the *first error* (FE), while when the last signal eventually disappeared we recorded the *last error* (LE), as illustrated in Figure 2.3-2. Measurements of the ion fluence needed to reach each of these two states were recorded and the SEFI FE and LE cross sections were calculated, being the chip area known.

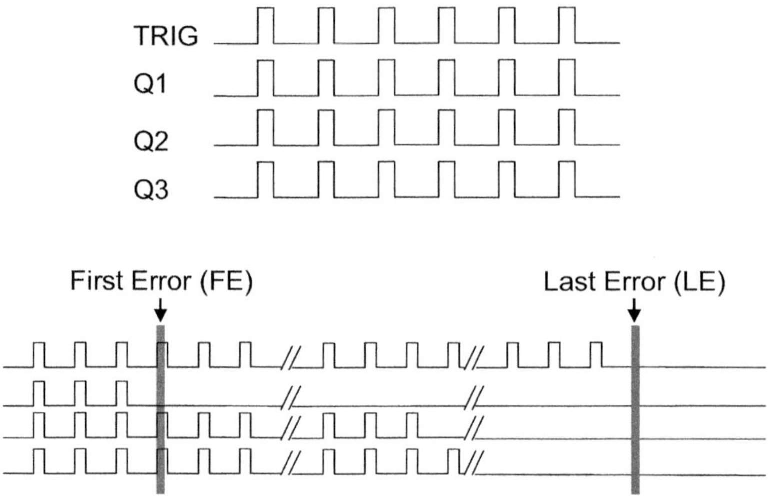

Figure 2.3-2. Output signals from the FPGA under test. TRIG is the reference signal and Q1, Q2, and Q3 are the control signals stating if the four SRs are working properly. First error occurs when one signal out of Q1, Q2, and Q3 is missing, while last error condition when the last signal fades.

In Figure 2.3-3 the statistical distribution of the experimental FE cross section, obtained from irradiation tests with Silicon ions, is plotted (over a logarithmic X-scale) and a Gaussian distribution appears as a reasonable fit to describe the data dispersion. Gaussian distribution fits also the other results obtained with ions listed in Table 2.3-1. By assuming a Gaussian distribution, we have estimated the confidence interval of each measured cross-section value, which abruptly changes for each exposure. The data variability depends on different factors, which are likely correlated with the reduced FPGA resources used by the design, that is, almost 20%. Together with the time to the *first error* also the error type was recorded by the control system. An examination of the error type is useful to outline the parts of the circuit more sensitive to SEUs (at least for the dynamic test). Figure 2.3-4 shows a histogram of the number of reference signals missing at the FE time. The results shown in Figure 2.3-4 are obtained from an average (normalized to 100 detected *first errors*) over all runs of several experiments. The sum of missing signals is not 100 because often the first error causes the disappearance of more than one signal.

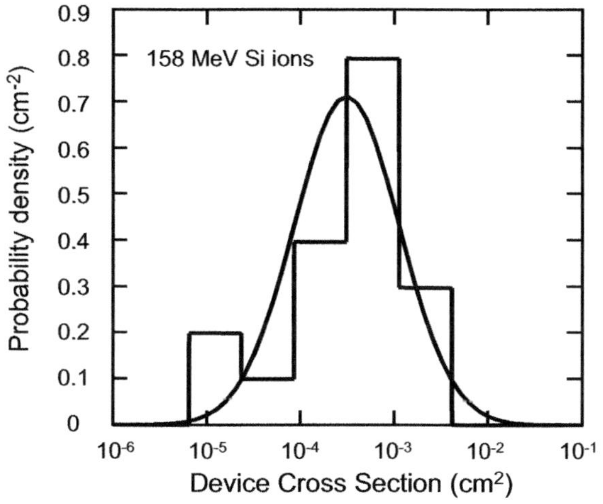

Figure 2.3-3. Statistical distribution of the experimental FE cross section obtained by irradiation with Silicon ions. For reference, we have plotted the best-fitting Gaussian distribution.

Q1, Q2, and Q3 represent the failing signals. A seemingly abnormal result is the high Q2 failure rate, being Q2 the result of comparison between two TMR SRs, which can be reasonably considered as SEU immune. This result suggests that SEFIs have to be correlated with the SEUs in the configuration memory that may induce a modification of the circuit implemented in the FPGA. Further, in this way it is demonstrated that the TMR technique is absolutely not efficient in reducing the SEFI in SRAM based FPGAs. On the contrary, the heartbeat signal (TRIG) almost never failed. This is reasonable, by considering that the circuit generating this signal is simple and consequently occupies a small area on the chip; further, it is controlled by few I/O lines.

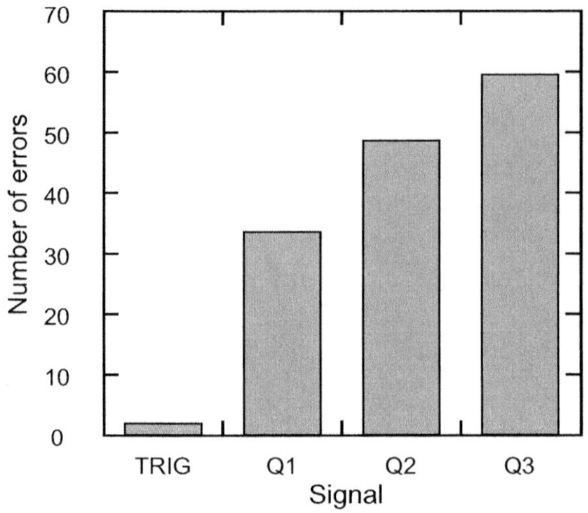

Figure 2.3-4. Number of times that the out signals disappear when the first error is detected. Data are normalized to 100 first error occurrences.

All SEFIs observed during the irradiation experiments did not induce any latch-up and were recoverable by cycling off/on the power. This suggests that SEFIs come from the corruption of the configuration memory and/or the modification of the lookup tables that implement the combinatorial logic in the DUT. In fact, a SEU in a RAM cell (either configuration or lookup bit) may cause the design corruption and hence the SEFI. In the FPGAs other elements exist, which are SEU sensitive, such as the JTAG TAP controller. These elements could induce SEFIs similar to those caused by the configuration memory/lookup table corruption, affecting as they do the configuration memory bits.

We can concentrate now on the *last error* and *first error* cross sections plotted in Figure 2.3-5 as a function of the ion LET.

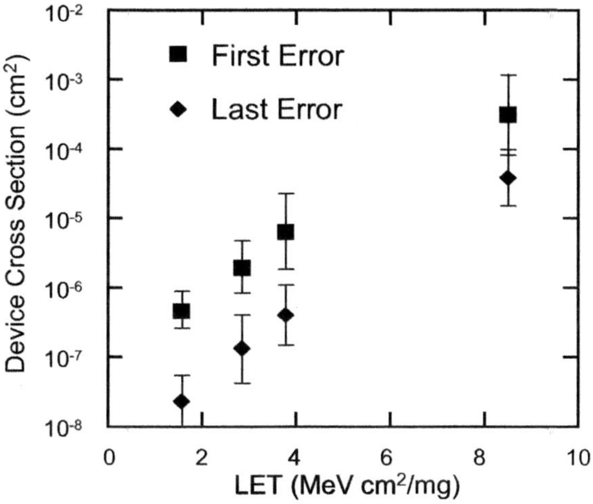

Figure 2.3-5. Device cross section for low LET values. Bars are twice the standard deviation.

The *first* and *last error* cross sections devices have been measured only for low LET ions (LET < 10) because for high LET ions the error rate is very high and our experimental setup could not properly measure the fluence to *first error*. Basically, this relates to the accurate measurements of very short time intervals, occurring from the electro-mechanical extraction of a faraday cup (i.e., the irradiation start) to the first system error.

3.2 Supply current increase: SEL?

Soft errors are accompanied by some changes of the DUT electrical characteristics due to the ion beam damage, which may even threaten the device integrity as in case of SEL. For this reason we monitored the occurrence of Icc≥800mA as discussed in Section 2. The fluence at Icc≥800 mA was recorded and the device cross section was calculated as in Figure 2.3-6. The data can be nicely fitted by the Weibull curve:

$$DCS(LET) = DCS_{sat}(1 - e^{-[\frac{LET - LET_0}{W}]^s}) \quad (1)$$

by using the parameter values: $DCS_{sat}=3.1\cdot 10^{-2}$ cm^2, $LET_0=0.1$ MeV cm^2/mg, W=32 and s=5.

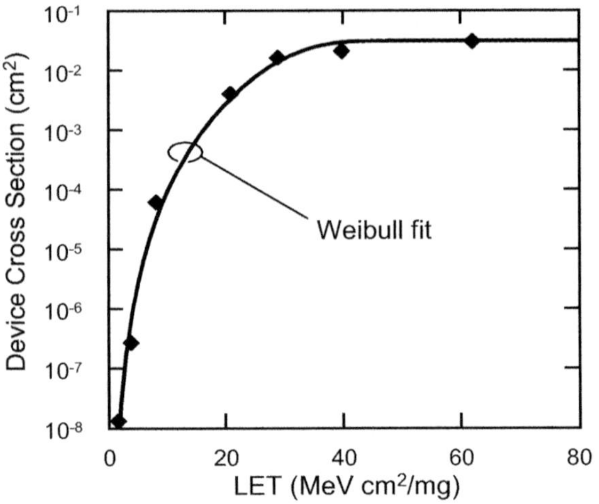

Figure 2.3-6. Device cross section relative to the condition Icc 800 mA. The solid line is the best fit to the Weibull curve.

Wondering if the 800 mA threshold represented a true SEL event, we compared the error type and the supply current during irradiation. Figure 2.3-7 shows the error code recorded by the control system during an irradiation run, while Figure 2.3-8 shows the corresponding supply current increase versus time during the same run. In Figure 2.3-7 we can see that when an error condition is reached, the state is stable for a long time and no different error codes are detected during this time. Namely, in Figure 2.3-7 the error code 9, standing for signals TRIG and Q1 correct and Q2 and Q3 always low, is recorded during the time period between 100 s and 1000 s after the irradiation start, an elapsed time that corresponds to a fluence of $3.2\cdot10^7$ O ions/cm^2. During this time period the supply current increased almost linearly with time and so with the dose (Figure 2.3-8). The error behavior of Figure 2.3-7 and the current increase of Figure 2.3-8 are not correlated. Hence, the current increase may be due to progressive SEU-induced driver contentions (e.g., two logic gates trying to impose different logic levels on the same output [WANG_99]) which are potentially high current absorption modes, but after a power off/on cycle the supply current returns at the original values (Icc=250 mA).

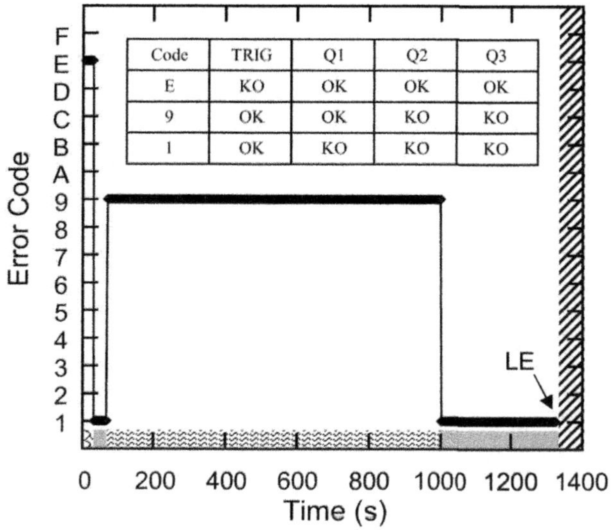

Figure 2.3-7. Code of the errors detected during irradiation with Oxygen ions.

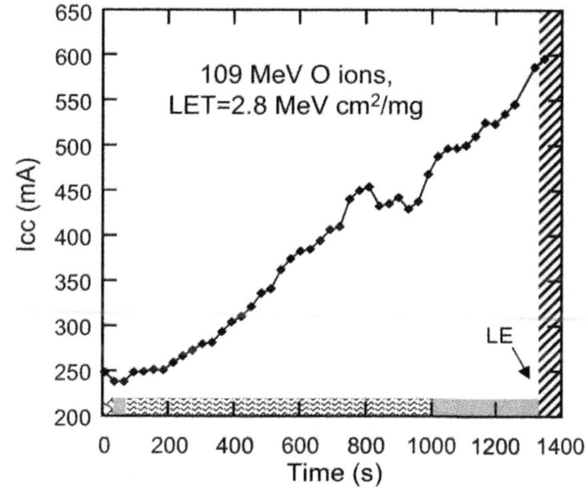

Figure 2.3-8. Icc current during irradiation with Oxygen ions measured during the same run of Figure 2.3-7.

A slow current increase during irradiation may be caused by a number of micro-latch-ups but not by a SEL, which hence is not observed in our devices. Unfortunately, distinguishing micro-latch from SEFI is not easy, because both of them induce a non-functionality of the circuit and an increase of the supply current [HOFF_00]. Another explanation could reside in the configuration memory corruption due to a non-functionality of the JTAG TAP controller that controls the configuration memory initialization. When a SEU in the memory elements of JTAG modifies the state, a large number of configuration memory bits could be changed in a short time. This event usually produces a sudden supply current increase, which has been detected few times during our experiments. Remarkably, the *last error* detection and the supply current increase at 800 mA occur almost at the same time, suggesting that they may originate from the same failure mechanism.

Finally, we performed a few irradiation runs after removing the Icc limitation at 800 mA. In this condition, the supply current was limited by the fold-back curve of a conventional power supply with maximum supplied current of 2 A. Still, these experiments never produced a permanent damage of the FPGAs.

3.3 SEU in the configuration memory

In the previous two sub-sections we have outlined that SEFIs are probably due to the corruption of the SRAM configuration memory. This assumption is corroborated by results found in [FULL_00], [KATZ_97], and [KATZ_98] who have performed radiation tests on Xilinx and Actel SRAM-based FPGAs. The investigations performed on Xilinx Virtex FPGAs were (partially) based on the read-back of the configuration bitstream. By using this methodology, Fuller et al. were able to compute with good precision the configuration cross section per bit and to locate the SEU corrupted bits in the memory. Altera FPGAs do not support this feature, so we cannot calculate exactly the cross section per bit. In order to obtain a coarse estimation of the cross section per bit, we have supposed that the *last error* measurements imply the corruption of almost all the configuration memory bits.

Results obtained by using the LE dynamic methodology are limited, unfortunately, only to low LET ions, due to the high SEE sensitivity of this component. However, we have underlined that the LE and Icc≥800mA cross sections likely derive from the same failure mechanism, suggesting the correlation between these phenomena. From this consideration and for simplicity we have assumed the same value for the two cross sections. This supposition is reinforced by the equivalence of the Icc value when the FPGA is unconfigured (roughly 800 mA) and the threshold value used in our experiment (Icc≥800 mA). In this way the configuration memory cross section has been also calculated and it is plotted in Figure 2.3-9.

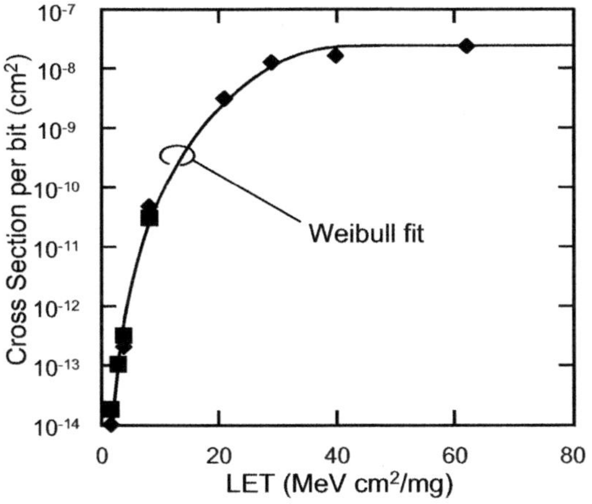

Figure 2.3-9. Global device cross section per bit data. ♦ Icc=800 mA, ■ last error.

4. CONCLUSIONS

We have presented heavy ion testing of SRAM-based Altera FPGAs. The choice of a design based on four shift registers has been dictated by the need for a simple yet efficient design to test all the different types of logic elements in the device. This test method has shown that SEFI errors are vastly dominant, because of the high SEU sensibility of both configuration memory and JTAG controller; in this sense, the choice of the particular design to implement in the DUT is irrelevant. The results found are in good accordance to what found in the literature for similar devices. A better comparison of the results and a thorough characterization of the device to heavy ions would imply the possibility of re-reading the configuration memory that, unfortunately, is not supported for these Altera devices.

PART 3

SOFTWARE-IMPLEMENTED FAULT INJECTION

Chapter 3.1

"BOND": AN AGENTS-BASED FAULT INJECTOR FOR WINDOWS NT

Andrea Baldini, Alfredo Benso and Paolo Prinetto

Politecnico di Torino, Dipartimento di Automatica e Informatica, Torino, Italy

Abstract: The goal of this chapter is to present BOND, a Software Fault Injection tool able to simulate abnormal behavior of a computer system running Windows NT 4.0/2000 Operating System. The Fault Injector is based on interposition techniques, which guarantees a low impact on the execution of the target program, and allows the injection of Commercial off-the-Shelf software programs. BOND allows performing both statistical and deterministic Fault Injection experiments, trading-off between overhead and precision of the obtained results. Moreover, the tool is capable of injecting faults into different locations, at any level of the application context (code and data sections, stack, heap, processor's registers, system calls...). A complete set of experimental results on different application programs demonstrates the effectiveness and the flexibility of the tool.

1. THE TARGET PLATFORM

Many studies and tools have been developed to address the problem of dependability validation under the UNIX operating system. Not the same can be said about the Windows NT operating system (OS), even if this OS is rapidly becoming the development, engineering, and enterprise platform of choice. Many Windows NT workstations are employed in enterprise-critical and even mission-critical applications, so the dependability of the Win32 platform (the platform which includes Windows NT) is becoming increasingly important. One significant example of the Windows NT trend is

A. Benso and P. Prinetto (eds.), Fault Injection Techniques and Tools for Embedded Systems Reliability Evaluation, 41-48.

the Information Technology in the 21st century (IT-21) directive [BIND_98], with which the U.S. Navy requires its ships to migrate to Windows NT workstations and servers. This also justifies the recent interest of the American Defense Advanced Research Project Agency (DARPA) in the field [GHOS_99]. Another example in specific Windows NT needs is the Italian most important aerospace company, Alenia Aereospazio, which is developing, in collaboration with our institution and the European Space Agency, a fault-tolerant PC platform running Windows NT to be used on the international space station Columbus.

Despite the proliferation of critical software for the Windows NT platform, usually the decision to use Win32-based systems is based on anecdotal evidence or at best on trust in the software vendor. This lack of scientific studies seems to be quite dangerous, also because it implies a lack of efficient methods to discover and correct dependable holes in a critical system. The choice of a commercial OS (so a COTS basic component) is a major disadvantage in the development of Fault Injection tools: since it is impossible to modify the source code of the OS, it is necessary to devise new and advanced techniques it in order to corrupt a particular application at the system or at the user level.

2. INTERPOSITION AGENTS AND FAULT INJECTION

Many modern operating systems use a system call-based interface between the OS and the applications. Intercepting these system calls and possibly modifying their behavior is a flexible way to add system functionality without accessing the original code or writing device drivers. This means that once individuated a common interface, or better once created an intermediate level between the OS and the applications, a specific agent can act as an extension of the OS for the application and as a wrapper separating the application from the OS (Figure 3.1-1).

Figure 3.1-1. The idea of Interposition Agents

Interposition agents can transparently record and possibly alter the communication occurring between the user code and the OS code. Both the target application and the OS run without noticing that the environment is slightly different from the original one. It is then straightforward that the idea of Interposition Agents can be exploited to implement an effective Fault Injection technique, which allows the injection and the monitoring of the target application with a low overhead and without requiring any modification neither of the injected program, nor of the OS.

3. THE BOND TOOL

The key objectives of the BOND tool are flexibility and stability, thus respecting the original idea of an automated tool for fault-tolerant validation under a significant number of experiments. Stability can be achieved with careful implementation and keeping the overall complexity as low as possible.

Flexibility should be considered in all the parts of the design, with particular attention to the adopted fault model, which is the argument of next section.

Different dimensions characterize the injection space, answering to four fundamental questions:

- When should the fault be activated?
- Where should the fault be injected?
- Which fault should be injected?
- How long should it last?

In the following, after describing the general structure of our Fault Injection environment, we will show how all these dimensions have been explored in order to provide a high degree of flexibility at the lowest possible cost in term of time, complexity and impact on the observed process.

3.1 General Architecture: the Multithreaded Injection

From a software-architecture point of view, BOND is implemented through two different Interposition Agents (Figure 3.1-2) that allow a very effective multithreaded injection:

- A Logger Agent, which is in charge of monitoring the target application in order to trigger the injection phase and observe the behavior of the target application after the injection.
- A Fault Injection Agent, which is in charge of actually injecting the selected fault into the target application.

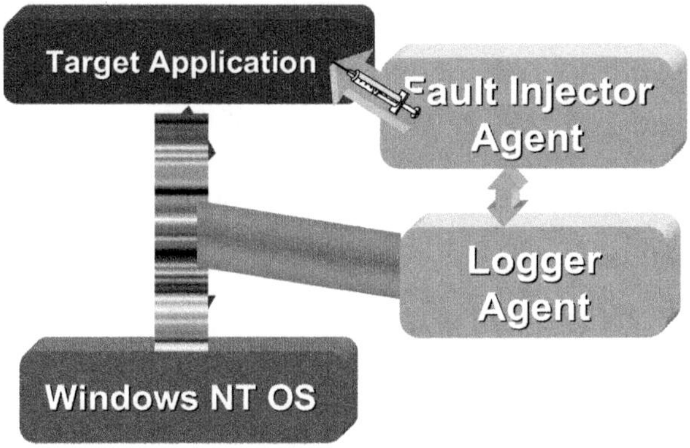

Figure 3.1-2. The BOND Structure

This structure allows on one side a first modularization, lowering the overall complexity of the injector, and adding flexibility to the approach. In particular, on a multiprocessor platform the two Agents could run in parallel, significantly reducing the execution time overhead of the target application. Moreover, in order to deal with commercial COTS software component, the tool fully supports the injection in multithreaded applications.

3.2 The Logger Agent

The Logger Agent is in charge of synchronizing the Fault Injector Agent with the target application and then observing the target application behavior after the actual injection.

The Logger Agent is able to detect the following three classes of events:

- Debug events. For each debug event notified by the OS kernel, the Agent records the target application context at the moment of the notification (execution time, process and thread IDs, processor's registers dump).
- API calls. For each API call (i.e., communication between the process and the OS), the Agent records the name of the called function, a list of its parameters and their type, an extract of the context (execution time, process and thread IDs, return address) and the return value of the function.
- Memory accesses. For each access to a particular region of memory (basically a list of sections), the Logger records the virtual address of the bytes addressed, the access type (read or write), the address of the instruction requesting the data, and an extract of the context (execution time, process and thread IDs). Every committed area of the process's memory (e.g., heap, stack) can be monitored.

Because of the variety of situations represented by all different COTS programs, the Logger Agent performs a preliminary complete mapping of the target process's memory. The individuation of the different regions and of their attributes is particularly important to decide the fault location in following injection phases. For each executable image (EXE or DLL loaded in process's memory) the injector identifies the different sections, with particular attention to the code and data sections. In addition, for each thread the injector finds the position and the limit (guard page) of the thread's stack. Finally, it is also able to locate the main heap of the process, along with other allocated regions of committed memory, both local to the process and shared with other processes.

Let's now detail the two main tasks of the Logger Agent, i.e., the Fault Injection Activation and the Fault Effects Observation.

3.2.1 Fault Injection Activation Event

The precision in the fault activation time is highly costly in term of time overhead, i.e., there is an evident trade-off between precision and speed. This is due to the conceptual program structure: a simple on-line observation allows low overhead but also a low precision, whereas a complex on-line observation typically implies a high monitored execution, at limits a step by step execution, with evident decrease of performances.

The BOND tool allows two different types of Fault Injection:

- *Statistical injection*:
  A non-deterministic injection, where the injection time is expressed as the elapsed execution-time from the beginning if the target application. The precision of the method is machine and program dependent. Obviously, the low intrusiveness (the only overhead is the observing process) is obtained at cost of a non-reproducibility of the exact experiment. This type of injection is particularly suitable for statistical studies over a large number of simple injections, where the speed of the single experiment is critical and the meaning of the injection campaign relies on the cardinality of the set of experiments.
- *Deterministic injection*:
  A reproducible experimental injection method, based on the count of some events (the limit is the count of the single instructions, in step-by-step execution). This method is expected to be quite expensive in terms of observation time, so it is suitable for complex injection campaigns with very precise experiments.

The first type of injection (statistical injection) is based on timers, whereas the second one (deterministic injection) needs a reliable synchronization with the Logger Agent, which provides on-line information about different counters of the logged events. The selected counting method has to be carefully selected: a uniform distribution of the monitored event allows a fixed measurement precision, whereas a non-constant distribution can lead to high precision but also to non-uniform Fault Injection experiments distribution over the target program execution time.

Because of the variability of the events' distribution with the single application, BOND performs this analysis before the beginning of the Fault Injection experiments, providing the user with a set of possible choices regarding the best suited "loggable" events.

Therefore, during Deterministic Fault Injection the Logger Agent allows to the Fault Injector Agent to work in two possible operative modes:

- API mode:
  The Fault Injection activation event is related with the count of a certain API call. Faults can be injected both before the requested call and after it, in order to allow modification of the function call's parameters and return values.
- Memory mode:
  The activation event is connected with a certain memory access (e.g. the 10th access to a certain memory area). The amount of monitored memory deeply influences the execution time overhead of the experiment. Virtually, the Logger Agent is able to monitor any area, but the

observation of the most critical areas only can drastically minimize the overhead.

3.2.2 Fault Effect Observation

After the injection of a fault, the Logger Agent observes the behavior of the target application. BOND classifies the result of each Fault Injection experiment in one of the following categories:

- No effect: the injected fault has no effect on the injected process, i.e., the current run is perfectly equivalent to a fault free run.
- Last chance notification: the injected program exits with a forced termination due to a non-recoverable exception.
- Silent fault (fail-silent violation): the injected program's execution seems correct, but there are differences between the actual results and the fault-free results. A silent fault is here considered a fault that creates no particular effect on program termination, but corrupts its final state, i.e., the results of the program. The Fault Injector Agent is able to detect this condition comparing a memory dump, i.e., a copy of the injected process's memory, obtained during a fault-free execution of the target program, and the one obtained during the actual injection. BOND allows the user to specify which memory area or even which single memory locations must be compared.
- Error messages: the OS rise a recoverable error and a consequent error message.
- Time-out: the injected program's execution is forcedly terminated because of a time-out condition, i.e., the time elapsed exceeds a given time-out value.

4. THE FAULT INJECTION AGENT

When activated by the Logger Agent, the Fault Injection Agent is in charge of actually injecting the next fault, selected from either an automatically generated or a user-defined fault list, into the target application. In the following we present the possible choices offered by the BOND tool in terms of fault location, type, and duration.

4.1 Fault location

As for the fault activation time, the flexibility in the choice of the fault location is crucial.

The presented approach allows injecting in every possible location in the application (process) context. In particular, two subspaces can be individuated, covering practically the entire process's context:

- Process's memory: the Fault Injector Agent can access to the low part of the process's virtual memory (lower 2 GB), that are reserved to user and DLLs' code and data. It is important to note that only the committed areas of the virtual memory are really allocated, whereas the non-committed areas (reserved and free regions) are not related to any physical memory and therefore can be not considered among the possible fault locations. The Agent is also able to inject faults into the process stack, giving the possibility to inject parameters and return values of high level functions, such as the API calls.
- Processor: the so-called thread context contains a copy of all the processor's registers, i.e., it contains the processor's state for that particular thread. Therefore, the modification of the context allows emulating faults appearing into the processor itself.

The combined possibilities of these location spaces can lead to the simulation of faults occurring in many different points of the system.

4.2 Fault type

Despite BOND supports the injection of different fault types, the Single Event Upset (SEU) has been chosen as the "default" fault for the experiments.

The question of how much this fault model represents real pathologies induced by the occurrence of real faults, is crucial [RODR_99]. Several software-implemented Fault Injection studies are dedicated to the analysis of the relationship between fault injected by software and physical faults.

The tool EMAX [KANA_93] has been developed to extract a realistic fault model from the observation of failures due to the occurrence of physical faults in a microprocessor. This study shows a large proportion of control flow errors affecting the address and data buffers, whose symptoms correspond, from an informational point of view, to the inversion of one or several bits. The work presented in [CZEC_93] shows that errors observed when injecting transient faults at the gate level in a simulation model of a processor can always be simulated by software at the informational level. The work described in [RIME_94] associates the occurrence of physical faults and errors at the functional level in a functional model of a microprocessor. This study concludes that almost 99% of the bit-flip faults can be emulated by software. Finally in [FUCH_98] the software Fault Injection is compared to ion radiation and physical injection, showing that a simple bit-flip model allows generating errors similar to those observed by physical injection.

All these studies tend thus to support the pertinence of the software-implemented Fault Injection technique based on bit(s) flipping at the target system informational level.

4.3 Fault duration

Using hardware terminology, the typical classification of faults divides the fault duration into three classes:

- Permanent faults: usually caused by irreversible device failures within a component due to damage, fatigue, or improper manufacturing.
- Transient faults: triggered by environmental disturbances (voltage fluctuations, electromagnetic interference, radiation). They usually have a short duration, returning to a normal operating state without causing lasting damage, but possibly corrupting the system state.
- Intermittent faults: they cause the system to oscillate between periods of erroneous activity and normal activity. They are usually caused by unstable hardware.

Again many researches [CARR_99_A], [CLAR_95] reveal that transient faults can be up to 100 times more frequent than permanents, depending on the system's operating environment, and they are much more significant in term of dependability simulation.

The Fault Injection tool presented in this paper allows the injection of both transient and intermittent faults. The injection of permanent faults, although possible, has not been implemented since it would introduce an unacceptable execution time overhead.

4.4 The Graphical User Interface

As shown in Figure 3.1-3, we implemented a very user-friendly interface, which allows the user to completely specify the Fault Injection campaign in terms of target program application, fault list, and statistical analysis of the Fault Injection results.

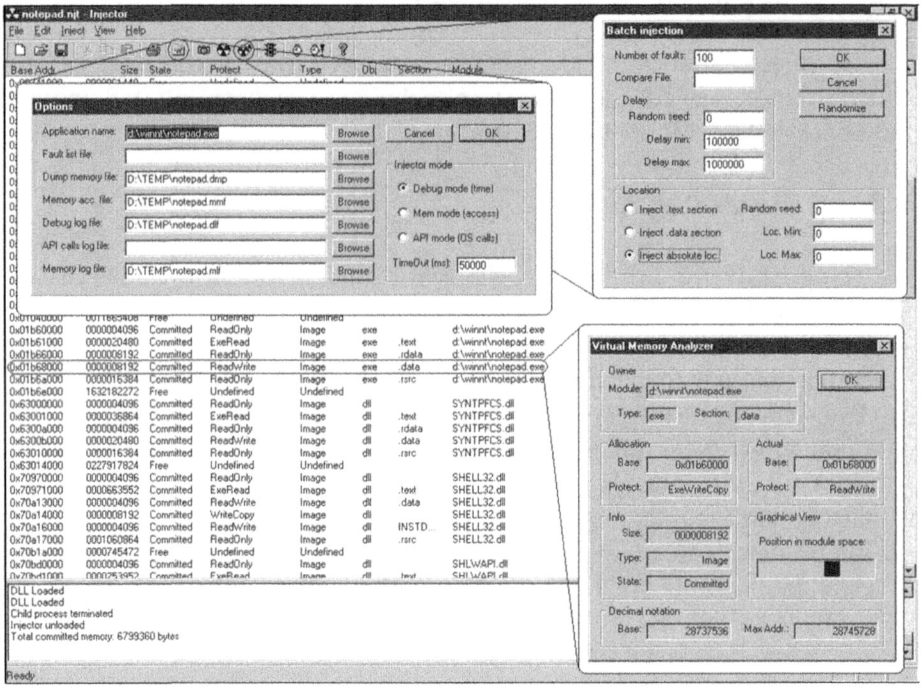

Figure 3.1-3. The BOND Graphical User Interface

5. EXPERIMENTAL EVALUATION OF BOND

In order to evaluate the effectiveness of BOND we performed different sets of experiments on custom and COTS software components (e.g., Excel 2000, Word 2000, WinZip, Notepad, and Calculator). In the following we present the two most representative set of experiments executed respectively on a widely used COTS application, Winzip32 7.0 for Win32 (Windows NT version), and a specific custom floating point benchmark application. The presented experiments have been executed using the *statistical injection* approach. We also repeated some of the experiments using the *deterministic injection* approach and, for a high number of injected faults (> 1,000) we noted the same results in terms of fault effects classification percentages, but an increased slow-down factor of at least 2 orders of magnitude. This result demonstrates, under particular assumptions, i.e., a high number of experiments, the reliability of the results obtained through the statistical injection.

5.1 Winzip32

Winzip32 is a widely used application used to compress files. Winzip32 is a much stable application, and it has a built-in support to detect and possibly correct errors (both at code level and at data level). These features can be used to show the power of the fault injector and of the detection capabilities of Winzip32.

The experiments are structured in seven independent injection campaigns, which refer to six sections of the injected process plus a "zoom" on the stack. The considered sections are the main exe data section (.data), main exe code section (.text), main heap, stack, first page of the stack, data section of compression routines (W32.data), and code section of compression routines (W32.text). During the experiment, the WinZip application compresses a single 1-MB bitmapped image. In each Fault Injection campaign we injected 5,000 bit-flip faults.

The fault effects shown in Table 3.1-1 have been classified in one of the following categories:

- NoEffect, i.e., the fault has no effects on he program behavior.
- LastChNot, i.e., last chance notification, corresponds to a forced termination of the application because of a critical error, and so the user can see the Winzip32 window suddenly disappearing, or often not appearing at all.
- FailSilViol, i.e., fail-silent violations, which are the most dangerous errors. In this case the application declares correct data generation, but the output file is corrupted.
- ToutNoEff, i.e., time-out condition without data creation. These cases correspond to Winzip32 detection, e.g. it appears a dialog box indicating data corruption or internal error.

Table 3.1-1. Fault Injection results in Winzip32

Section	NoEffect	LastChNot	FailSilViol	TOutNoEff
.data	99.5 %	0.0 %	0.0 %	0.5 %
.text	91.3 %	1.3 %	0.3 %	7.1 %
Heap	89.2 %	2.1 %	0.0 %	8.7 %
Stack	99.8 %	0.1 %	0.1 %	0.0 %
Stack 1st page	99.6 %	0.2 %	0.2 %	0.0 %
W32 .data	92.3 %	0.7 %	3.7 %	3.3 %
W32 .text	90.0 %	1.1 %	0.4 %	8.5 %

The FailSilViol result is particularly low (3.7 %), also because in many other cases (3.3 %) the detection mechanism correctly warns the user about the fault. It is also significant that in the same case the 92.3 % of injections have no effect, so they are either in a non-critical area of the section or they are corrected by the application.

It is also interesting to note that the detection mechanism works particularly well with injection into code sections and heap, and that faults injected into stack have little effect, showing that high calculation complexity does not always affect the use of the stack.

5.2 Floating Point Benchmark

This application is a custom software program that perform many (over one million) floating point operations over a set of 1024 "double" (64-bit) variables. The benchmark is in two versions, one "original version" and one modified "fault-tolerant version". The program outputs a memory dump of the variables at the end of the computation. This second version performs internal checking using variable reordering and duplication, trying to detect data errors.

The experiments are structured in five independent injection campaigns for the original version (Table 3.1-2), plus other five independent injection campaigns for the fault-tolerant version (Table 3.1-3), to demonstrate the flexibility of the fault-injector and the effectiveness of the fault effect classification.

The considered sections are main exe data section (.data), main exe data section limiting the injection to the effective addresses of the variables (.data variables), stack, main exe code section (.text), first page of main exe code section (.text 1st page). Again, in each Fault Injection campaign we injected 5,000 bit-flip faults.

As expected, the fault-tolerant version of the benchmark presents a considerably reduction of the FailSilViol cases.

Table 3.1-2. Fault Injection results in the original floating point benchmark

Section	NoEffect	LastChNot	FailSilViol	TOutNoEff
.data	77.8	4.3	17.9	0.0
.data variables	40.9	0.0	59.1	0.0
Stack	99.3	0.5	0.2	0.0
.text	92.1	3.6	4.3	0.1
.text 1st page	13.3	46.5	39.2	0.4

Table 3.1-3. Fault Injection results in the Fault-Tolerant floating point benchmark

Section	NoEffect	LastChNot	FailSilViol	TOutNoEff
.data	97.0 %	3.0 %	0.0 %	0.0 %
.data var	99.3 %	0.0 %	0.7 %	0.0 %
Stack	99.6 %	0.4 %	0.0 %	0.0 %

.text	94.0 %	4.6 %	1.3 %	0.1 %
.text 1st page	96.9 %	1.7 %	1.4 %	0.0 %

6. CONCLUSIONS

In this chapter we presented BOND, a powerful Fault Injection tool running under Microsoft Windows NT 4.0 OS. Exploiting the idea of Interposition Agents, BOND is able to inject faults in COTS software applications without requiring any modification of either the OS or the target application itself. BOND allows performing both statistical and deterministic Fault Injection experiments, trading-off between overhead and precision of the obtained results. Moreover, the tool is capable of injecting faults into different locations, at any level of the application context (code and data sections, stack, heap, processor's registers, system calls,...). A complete set of experimental results on a COTS application and on a custom benchmark has been presented to demonstrate the effectiveness and the flexibility of the tool.

Chapter 3.2

XCEPTION™ : A SOFTWARE IMPLEMENTED FAULT INJECTION TOOL

Diamantino Costa[2], Henrique Madeira[1], Joao Carreira[2] and Joao Gabriel Silva[1]

[1] *University of Coimbra, Portugal*
[2] *Critical Software SA*

Abstract: This chapter addresses Xception - a software implemented fault injection tool. Among its key features are the usage of the advanced debugging and performance monitoring resources available in modern processors to emulate realistic faults by software, and to monitor the activation of the faults and their impact on the target system behaviour in detail. Xception has been used extensively on the field and is one the very few fault injection tools commercially available and supported.

Key words: Fault-injection, Software Implemented Fault-Injection, Dependability Evaluation

1. INTRODUCTION

Software Implemented Fault injection (SWIFI) is probably the only fault injection technique that is not significantly affected by the increasing complexity of processors. The basic idea of SWIFI consists of interrupting the application under execution in some way and executing specific fault injection code that emulates hardware faults by inserting errors in different parts of the system such as the processor registers, the memory, or the application code.

The Xception project used a hybrid approach to get the flexibility without the overhead, by relying on the advanced debugging and performance

A. Benso and P. Prinetto (eds.), Fault Injection Techniques and Tools for Embedded Systems Reliability Evaluation, 41-48.

monitoring features existing in modern processors to inject more realistic faults by software and to monitor the activation of the faults and their impact on the target system behaviour in detail. This approach has become a de facto standard for later SWIFI tools, like MAFALDA [Chapter 3.3], NFTAFE [STOT_00], and GOOFI [AIDE_01].

Xception has been used in various domains, ranging from Online Transaction Processing Systems [COST_00] to COTS system for space applications [MADE_02]. The supported target processors include PowerPC, Intel Pentium and SPARC based platforms running WindowsNT/2000 and Linux OSs and several real-time systems such as LynxOS, SMX, RTLinux, and ORK.

The structure of this chapter is as follows. Section 2 describes the Xception approach and its relation with related research. Section 3 addresses the Xception toolset, describing the architecture and key features and presenting in brief a case study using the Xception tool. Section 4 concludes with a critical analysis of the SWIFI approach and the Xception tool.

2. THE XCEPTION TECHNIQUE

Xception uses the advanced debugging and performance monitoring features existing in modern processors to inject realistic faults by software, and to monitor the activation of the faults and their impact on the target system behaviour in detail. The breakpoint registers available in the processors allow the definition of many fault triggers, including fault triggers related to the manipulation of data. The processor is running at full speed and the injection of a fault corresponds normally to the execution of a small exception routine. On the other hand, by using the performance monitoring hardware inside the processor, detailed information on the behaviour of the system after the injection of a fault can be recorded. Some examples are the number of clock cycles, the number of memory read and write cycles, and instructions executed (including specific information on instructions such as branches and floating point instructions) from the injection of the fault until some other subsequent event, for instance the detection of an error (latency).

The combination of the trigger mechanisms provided by the debugging hardware and the performance monitoring features of the processor, allows monitoring other aspects of the target behaviour after the fault with minimal intrusion. For example, it is possible to detect if a given memory cell was accessed after the fault or if some program function (e.g., error recovery routine) was executed. Another important aspect is that, because Xception operates very close to the hardware (at the exception handler level), the injected faults can affect any process running on the target system including the kernel. It is also possible to inject faults in applications for which the source code is not available.

2.1 The FARM model in Xception

The FARM model introduced in [ARLA_90] defines the key attributes required to characterize a fault injection process. The model considers an input domain including the set of faults F and the activations A, used to functionally exercise the target system, and an output domain composed by the readouts R and a set of measures M computed from the readouts.

2.1.1 Faults

Table 3.2-1 show the elements required to characterize a fault in the Xception approach. Xception faults can directly emulate physical transient faults affecting many internal target processor units, main memory, and other peripheral devices that can be accessed by software. The fault triggers are based on breakpoint registers and allow the injection of faults in practically any circumstances related to code execution, data manipulation, or timing. The fault types consist of bit manipulations, which is a widely accepted model for hardware faults.

Table 3.2-1. Xception main fault parameters

Fault Locations	Fault Triggers	Fault Types
Processor Integer Unit Floating Point Unit Memory Management Unit Internal Data Bus Internal Address Bus General Purpose Registers Branch Processing Unit Processor cache (partially) System Main memory Devices accessible by software	Opcode fetch from a specified address Operand load from a specified address Operand store to a specified address After a specified time since start-up A combination of the above	Bit level operations: stuck-at-zero stuck-at-one bit flip. Bit mask (32 bits) Number of bits to affect

The injection of a fault in Xception includes several steps. The very first step (before the actual injection of the fault) is to program the fault trigger in the processor debugging logic. The injection run starts after this point and the target system runs at full speed. When the exception that injects the fault is activated, specific registers or memory content are corrupted according to the fault type and functional unit to be affected by the fault. Depending on the target functional unit and trigger type, a third step may be required to restore the original memory/register contents. This last step is necessary in several situations to guarantee that real transient faults are emulated.

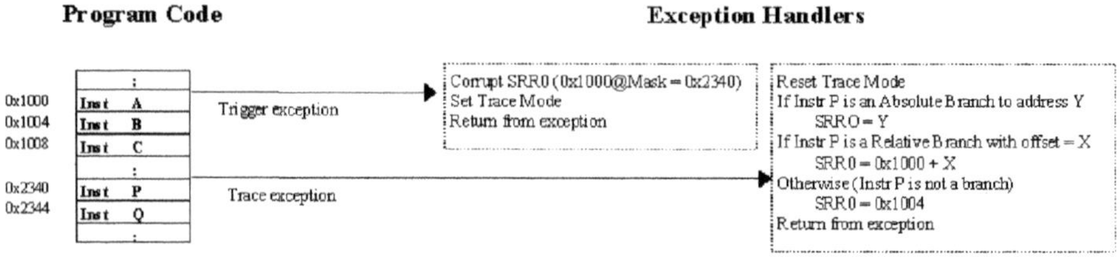

Figure 3.2-1. Transient address bus error when an opcode is fetched (PowerPC)

As an example, Figure 3.2-1 illustrates the main steps needed to inject a fault on the address bus when a code is being fetched in a PowerPC7xx processor. When the trigger instruction (Inst. A) is reached and Xception gains control over the processor, the register SRR0 is corrupted. Execution is then allowed to resume in Trace mode at the corrupted address. If no serious error occurs, such as an access to an invalid address, a Trace exception is raised immediately (after the execution of the instruction at the corrupted address). Xception then determine the address where the program has to resume the execution and put this value in SRR0. This address is dependent on the type of erroneous instruction executed as illustrated in Figure 3.2-1. If the instruction used an absolute branch, then the branch is taken normally. If it was a relative branch, it is also taken, but the offset is relative to the original instruction (A). Finally, if it was not a branch instruction, execution continues in the instruction immediately after A.

Each fault type requires a specific algorithm for its emulation but all of them follow steps similar to those presented in Figure 3.2-1.

2.1.2 Activations

The activations in the FARM model represent mainly the workload executed in the target system and the operational profile of the workload. Unlike other tools, Xception does not put any restrictions to the activations set. Source code is not required (i.e., any program can be used) and, as the faults can be injected in both user and kernel space, all the code running in the system can be considered for the activation set.

2.1.3 Readouts

Table 3.2-2 show the different types of readouts that can be collected by Xception. Some readouts are, obviously, dependent on the target system.

It is worth noting that only the injection context has to be collected during the critical period when the fault is injected. All the other readout types are collected after stopping the injection experiment to reduce the interference to the minimum possible.

2.1.4 Measures

The measures are very dependent on ability of the tool to inject relevant faults and to collect the required readouts. From the readouts collected by Xception it is possible to compute directly common measures such as failure modes, and measures related to the efficiency of error detection mechanisms (coverage and latency), recovery mechanisms (efficiency and recovery time), and other fault tolerance mechanisms available in the target system ([COST_00], [MADE_02]). Other types of measures are normally dependent on the specific experiment.

Table 3.2-2. Xception Readouts

Readout type	Description
Fault parameters	Fault trigger, type, and location.
Injection context	Processor state at time of fault: instruction pointer, opcode of the affected instruction, register state, etc.
Execution time	Execution time of the application measured from the Xception host.
Exit code	Application' s exit code (e.g. returned by the exit() call in the C Language).
Operation system errors	Error detected and reported by the OS.
Application(s) output correctness	Result of comparison of the program output with the correct outputs.
Error detection mechanisms	Errors detected by existent Error Detection Mechanisms (EDM) in the target, including detection latency.
Monitoring data	Reserved for data collected from the processor performance monitoring hardware

3. THE XCEPTION TOOLSET

The Xception™ product family includes several tools:

- The original Xception™ tool, based on Software Implemented Fault Injection (SWIFI) technology.

- The Easy Fault Definition (EFD) and Xception Analysis (Xtract) add-on tools.
- The extended Xception™ (eXception) tool, with the fault injection extensions based on scan chain technology and pin-level forcing technology.

All tools share a common and key component: a graphic user interface named Experiment Manager Environment (EME), providing the backbone for the whole family of tools, thus offering a consistent view to the user.

3.1 Architecture and key features

The Xception™ architecture (Figure 3.2-2) resembles the Client/Server model. It comprises a front-end module (EME), which runs in a host computer and is responsible for experiment management/control, and a lightweight injec-tion core, which runs in the sys-tem under evaluation and is responsible for the insertion of the faults.

The connection between the host and the target is done by means of a high level protocol, built on top of TCP-IP. Experiment and fault configuration data flows from the host to the target, while fault injection raw data results flow in the opposite direction.

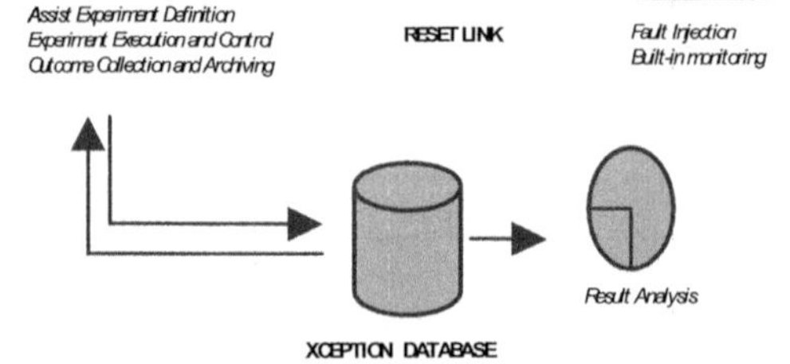

Figure 3.2-2. The Xception Architecture

A specific reset "link" is provided to enable restart of the target system, in case it becomes corrupted, frozen or crashes due to a fault, so that a new injection run can be started in a clean state, thus ensuring a fully automated and unattended experiment execution. This reset link is operated from the host machine through a standard serial port, on which one of the handshake signals is used to connect to an hardware reset input line on the target systems.

3.1.1 The Experiment Manager Environment (EME)

The Xception Experiment Manager Environment (EME) runs in the host computer and is responsible for fault definition, experiment execution and control, outcome collection and basic result analysis.

Xception uses a SQL relational database, enabling extensive outcome analysis. The Xception database holds all the information relevant to the fault injection process, including fault definition, experiment execution and result analysis. The use of an open standard such as SQL also allows the user to execute specific SQL queries or use other tools available in the market to explore the results (in addition to the tool already provided by the Xception).

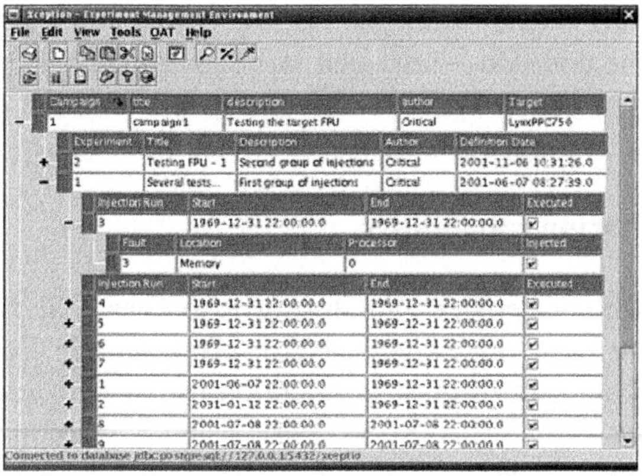

Figure 3.2-3. The Experiment Manager Environment (main window)

3.1.2 On the target side

The Xception Injection Run Controller is a key component that includes the algorithms that enable Xception to emulate faults and perform the required low-level synchronization with the target system.

The Injection Run Controller holds a lightweight injection core, installed in the target system at the CPU exception handling level (the reason for the name 'Xception' for the tool), performing fault injection through software-only manipulations.

Xception uses the built-in debugging features of contemporary processors to inject faults with minimum intrusiveness. Fault triggers are implemented using the processor breakpoint registers and therefore the system is allowed to run at full-speed. During "normal" system operation, fault-injection code in the target behaves as "dead code". For most of the fault models only a single instruction runs in trace mode, which makes Xception almost non-intrusive. For instance, on an Intel PentiumIII at 800 MHz the fault-injection overhead is typically below 0.85 μsec for the Xception Linux product.

Concerning memory overhead, the Xception injection kernel occupies as little as 30 Kbytes.

Besides the injection core, there is a Communication Handler, which wraps all communications with the host side, collecting fault parameters at start-up, and routing the fault-injection results back to the host in the end.

The user may freely choose the program to run in the target system during fault injection (the "workload"). The source code does not need to be available, and the code does not need to be instrumented (modified). Faults may be injected in user-level applications as well as in the operating system (kernel included), assuring a truly system-wide coverage.

3.1.3 Monitoring capabilities

Xception offers extensive monitoring resources that are invaluable to track odd faults and investigate error propagation. Besides processor performance monitoring resources, Xception provides extensive log support that might be used, for instance, to assist in the validation or evaluation of fault-tolerant mechanisms. Additionally, an entry point in the Xception target support library (in the form of a registered callback) enables system evaluators and testers to log any events and data they want. By default a snapshot of the processor is logged at the injection instant and on selected error detection events (e.g. panic routines of the OS or exceptions not handled). The level of detail used for logging fault-injection events can be decided/configured by the user. However, in spite of the use of low intrusive techniques based on the processor breakpoint registers and performance counters, some overhead can be caused if fine detail logging is needed.

3.1.4 Designed for portability

Xception deployment for a specific target/fault model requires developing/porting the following modules:

- the specific plug-in for the EME.
- the specific fault injection logic.
- the target specific communication handler

The effort put on an instantiation of the Xception tool for a particular usage/platform depends essentially on the two following aspects:

1. on the level of architectural platform differences on a new to-be-supported target system (against the existent supported platforms);
2. on the novelty of the fault model (when compared to the already supported fault models).

3.2 Extended Xception

The extended Xception suite of products expands the Xception by incorporating non-SWIFI fault-injection technology. For non processor-centric designs, as found in Telecommunications, Networking and Automotive Embedded solutions, direct fault placement in Application Specific Integrated Circuits (ASICs), chipset devices or even nodes in a PCB (Printed Circuit Board) may be a requirement. Even when considering internal processor locations, standard software access does not reach all functional units - caches being one good example of this limitation.

The extended Xception [SANT_01] approach relies on a modular architecture that uses the same experiment definition and control front-ends (the Xception Experiment Management Environment) and the same database with two new hardware modules. These modules deliver pin-level forcing injection and scan chain implemented fault injection (SCIFI). Each technique - pin-level injection, SCIFI or SWIFI - is totally independent from the others and it may be used alone or concurrently, depending on the evaluation goals and features of the target system under evaluation.

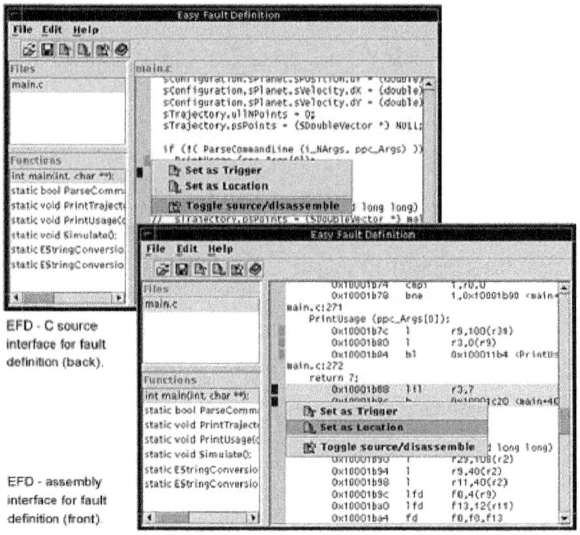

Figure 3.2-4. The Easy Fault Definition tool

3.3 Fault definition made easy

The Easy Fault Definition (EFD) is an add-on tool enhancement to the Xception's front-end application aimed at simplifying considerably the operator effort put on experiment definition (i.e. assisting mainly fault and trigger definition).

The EFD tool allows the user to browse through an application's source code and interactively mark memory ranges, within the code and data spaces, as either fault locations or fault triggers. This is achieved by using a very simple interface, and with only a few mouse clicks or keystrokes.

In case the source is not available, or the user wants to target a specific machine code instruction, translation to machine level code instructions is done on the fly, and the same browse and interaction operations are also available.

3.4 Xtract – the analysis tool

The Xtract add-on tool provides pre-defined queries onto the Xception database, whose results are presented to the user in an easy-to-understand layout. It also provides the possibility to exhibit different kinds of statistical graphics, according to the query results.

With Xtract, the user is allowed to browse the four main entities in the Xception database: Campaigns, Experiments, Injection Runs and Faults. Besides all the information provided on each specific entity, the Injection Runs and Faults entities also come with its associated outcomes: the workload exit code, classification, trigger and target addresses.

The Xtract tool may also be used to carry out fault/error impact analysis using an *Easy Fault Definition* like display, i.e., displaying the fault impact at the source code abstraction level. To do this, the user simply needs to configure the Xtract tool with the list of modules to be inspected - all the memory and symbol mapping is done in a transparent way to the end user.

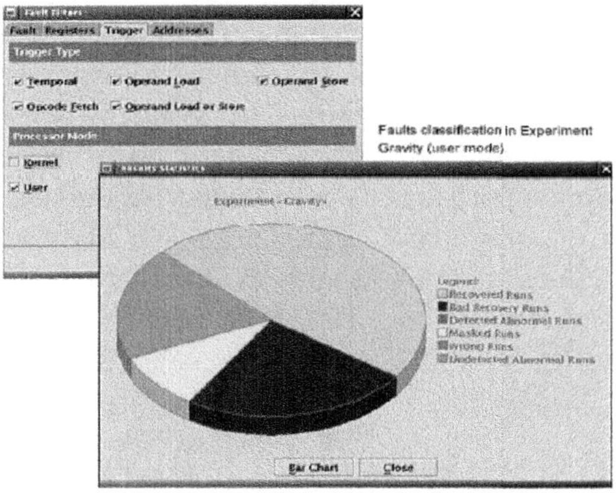

Figure 3.2-5. A detail of the Xtract tool

3.5 Xception™ on the field – a selected case study

This section provides a brief overview of a recent experimental study performed using the Xception SWIFI tool. It was done together with NASA's Jet Propulsion Laboratory and University of California, Los Angeles (UCLA), on a prototype of a COTS-based system for space applications (mainly scientific data processing) chosen in the Remote Exploration and Experimentation (REE) project [MADE_02]. The objective was to evaluate the impact of transient errors in the operating system of a COTS-based system, quantifying their effects at both the OS and the application level.

The total effort was about two man-months, and involved many iterations with new injection campaigns being devised as data from experiments was analysed. The capabilities of Xception for easy definition and unattended execution of fault-injection experiments were intensively used.

3.5.1 Experimental setup

The target system was a CETIA board with two PowerPC 750 processors and 128 Mbytes of memory, running Lynx operating system version 3.0.1. The host was a Sun UltraSparc-II with SunOS 5.5.1. The target system was restarted after each injection to assure independent experiments. Faults were injected after the workload start and, depending on the type of trigger, faults were uniformly distributed over time or were injected during the execution of specific portions of code.

A synthetic workload was defined in order to exercise core functions of the operating system such as the ones related to processes (schedule, create, kill, process wait), memory (allocate memory to a process, free memory), and input/output (open, read, write). The synthetic workload executed a given number of iterations, and in each iteration it started by doing some buffer and matrix manipulations to use memory resources that were checked for integrity later on, and then executed a number of system calls related to the core OS functions mentioned above. Three instances of this synthetic program were used (P1, P2, and P3), each with a different resource usage profile (Figure 3.2-6).

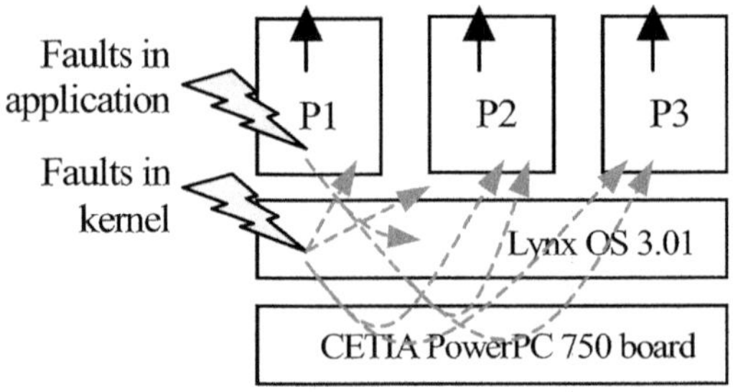

Figure 3.2-6. Processes configuration for the synthetic workload

3.5.2 Results

The full results from this research encompass a comprehensive picture of the impact of faults on LynxOS key features (process scheduling and the most frequent system calls), data integrity, error propagation, application termination, and correctness of application results [MADE_02]. This section outlines only a specific part of the results, regarding error propagation among user level processes, to give an idea of what Xception can achive. For this purpose, faults were injected only while process P1 was scheduled, and the effects of those faults in the other two processes were monitored. Table 3.2-3 shows the failure mode classification used:

Table 3.2-3. Failure mode classification

System crash – All the processes crashed: OS has to be rebooted.	**No impact** - All the applications terminated normally and produced correct results
Application damage - Fault damages were confined to P1 (other processes executed normally). The following damages are considered: **Application crash** - The process P1 (the target application) crashed. No results have been produced by P1. **Errors detected** - Errors have been detected at the app. level (e.g., mem. consist. checks, error codes ret. by OS, etc) **Wrong results**: The application terminated normally (no errors detected) but produced incorrect results.	**Error propagation** - Faults affected at least one of the other processes but the system did not crash. The following propagation types apply: **Other application crash**: The application crashed. No results have been produced by the application. **Errors detected in other application**: Errors have been detected at the application level. **Wrong results in other application**: Other application terminated normally (no errors detected) but produced incorrect results.

A total of 975 application faults were injected when P1 was executing in user mode (i.e. application code), and 1038 OS faults were injected when P1 was executing in kernel mode (i.e. OS code).

The percentage of faults whose damage was confined to P1 is very high. Only 53 faults resulted in error propagation, and of these only 3 application faults and 6 OS faults escaped detection and caused the other applications (P2 or P3) to produce incorrect results. This means that LynxOS does a good (but incomplete) job in confining the errors to the process that is directly affected by the fault. In demanding applications further isolation measures may be needed. The full results regarding error propagation are presented in Table 3.2-4.

Table 3.2-4. Error propagation details

Error propagation detail	**Number of faults**	
	Kernel mode	User mode
Other application crash	2	4
Errors detected in other applications	2	36
Wrong results in other applications	6	3

Errors detected in other application		# faults
Appl. Level	Memory corruption	11
OS level	SIGTRAP ' (trace mode)	5
	SIGPIPE (write + no pipe read)	2
	SIGSYS (bad arg. to system call)	18

4. CRITICAL ANALYSIS

4.1 Deployment and development time

One of the main advantages of SWIFI is the ease with which a new fault injection campaign can be started. Two connections between the host and target are needed: a link to exchange messages, and a line that enables the host to reset the target (this may be the most difficult part because this connection is not standard). Then a piece of code has to be installed in the target, the main tool in the host, and optionally some calls to the logging subsystem inserted in the application to enhance data collection. A knowledgeable user should take only a day.

If no plug-in exists for a particular architecture, then indeed it will take much longer. If only the target processor is new, then a new plug-in should take a couple of months for one person to develop. If the target operating system is also new, then the effort depends strongly on the availability of source code. It should take an additional month for an open OS, but the reverse engineering required to make the target module of the tool coexist with a closed OS can take much longer.

The target independent part of the Xception took about three man years to develop, not counting the previous experience in stabilizing the specification. This is a commercial product; a simpler experimental tool for research should only take three to five man-months to develop.

4.2 Technical limitations of SWIFI and Xception

SWIFI does have some intrusiveness. If the processor has reasonable breakpoint capabilities, this overhead can be very small for transient and intermittent faults, both in time and memory, but it is not null. Permanent faults require single step execution, which has a severe overhead, except for those special cases where the hardware triggers of the breakpoint mechanism enable a more efficient implementation, like a stuck-at fault in memory forced by a breakpoint on every access to that memory cell. Xception does not support permanent faults.

Being a software implemented technique, SWIFI does not allow the insertion of faults in non software accessible parts of the processor, like pipeline control and hidden registers, and the computer at large, like I/O circuits. Xception circumvents this problem by emulating the consequences of faults that occur in any such place (i.e. the resulting errors), but the models required can be hard to devise, essentially because the detailed architecture of VLSI circuits is rarely available, or because of the high cost of their development. Adding a model to emulate faults e.g. in a particular I/O subsystem requires changes, in the case of the Xception tool, to the plug-in specific to the target architecture.

The requirement that a remote reset link be available can be also an hurdle, not too difficult to solve, but inconvenient.

SWIFI gives no help in determining fault rates in each subsystem of a computer. These have to be obtained from other sources. It can only be used to study the consequences of faults, should these occur. Still, a problem arises when deciding how to spread a fault-injection campaign over many sub-systems, to get aggregate figures like coverage. Simple models can be used, like injecting in each sub-system a number of faults proportional to the silicon area occupied, maybe compounded by the duty cycle of that sub-system. Since research has not yet given any clear answer to this problem, Xception gives no help in this respect.

ACKNOWLEDGEMENTS

The authors are very grateful to all the people that have been contributing to the enhancement and constant evolution of the Xception tool. Special thanks to António Alves, Francisco Moreira, Luis Henriques, Nuno Duro, Telmo Menezes, and Ricardo Maia at Critical Software SA, to João Cunha at the ISEC, to people at LAAS, France, Chalmers, Sweden, and Urbana-Champaign, Illinois, and last but not least a special thanks to the anonymous referees that have reviewed our work for the precious comments that helped us keeping improving the Xception toolset.

Chapter 3.3

MAFALDA: A SERIES OF PROTOTYPE TOOLS FOR THE ASSESSMENT OF REAL TIME COTS MICROKERNEL-BASED SYSTEMS

Jean Arlat, Jean-Charles Fabre, Manuel Rodríguez and Frédéric Salles

LAAS-CNRS, Toulouse, France

Abstract: MAFALDA (Microkernel Assessment by Fault injection AnaLysis and Design Aid) encompasses a series of prototype tools providing quantitative information on real-time COTS microkernels to support their integration into systems with strict dependability requirements. We illustrate how the most recent version of MAFALDA, namely MAFALDA-RT, is organized, the basic fault injection techniques it implements, the main experimental parameters that are to be specified, and the various measures that can be obtained. Finally, we draw the main lessons learnt and some perspectives for this work.

Key words: Real-time microkernels, COTS components, software-implemented fault injection, robustness assessment, characterization of failure modes.

1. INTRODUCTION

The new trend for building computer systems relies on the use of COTS (*Commercial-Off-The-Shelf*) software components. The main raisons for this tendency are motivated by development cost and time-to-market reductions [BOEH_99]. Nevertheless, the decision whether to use COTS microkernels into embedded systems still raises serious problems from a dependability viewpoint [VOAS_98], in particular for safety-critical applications (e.g., aerospace and railways systems).

The original MAFALDA prototype tool has been developed to address this problem. The first goal is to provide means for objectively

A. Benso and P. Prinetto (eds.), Fault Injection Techniques and Tools for Embedded Systems Reliability Evaluation, 41-48.

characterizing the behavior of a microkernel in the presence of faults. The second important feature of MAFALDA concerns the support offered for implementing error confinement wrappers that are aimed at improving the failure modes exhibited by the target microkernel [ARLA_02]. To the best of our knowledge, none of the existing tools, like FERRARI [KAOI_93], FINE [KANA_95], Xception [CARR_98], [Chapter 3.2] and Ballista [KOOP_99], support such a double aim of i) analysis by fault injection and ii) error confinement by wrappers. The MAFALDA environment (e.g., see [ARLA_02]) distinguishes from them by: i) the range of faults injected, ii) the broad spectrum of analyses allowed, and iii) the facilities provided to specify, design and implement wrappers [SALL_99] [RODR_00] [RODR_02_A].

The most recent version MAFALDA-RT (*Microkernel Assessment by Fault injection AnaLysis for Design Aid of Real Time systems*) generalizes these capabilities and extends them to cope with the specific requirements of real-time systems [RODR_02_B] [RODR_02_C]. The dual goal of assessing the robustness and of improving the dependability properties of a target commercial microkernel are addressed by i) characterizing its failure modes in both the time and the value domain, and ii) hardening it with error confinement wrappers. The first goal is undertaken by using *Software-Implemented Fault Injection* (*SWIFI*) targeting faults both external and internal to the microkernel. The second goal is achieved by means of *wrappers* derived from properties expressed in temporal logic; these wrappers concurrently check the behavior of the microkernel functional components (synchronization, scheduling, timing, etc.) from specifications expressed in temporal logic. These wrapping features are implemented with a reflective approach that provides the necessary levels of observability and controllability [MAES_87].

For sake of conciseness and to better fit with the scope of the book, in the sequel of this chapter we concentrate on the robustness assessment capabilities of MAFALDA-RT. In particular, the tool supports two novel features: i) virtual elimination of temporal intrusiveness and ii) enhancement of the observations in the temporal domain. Another major improvement concerns the management of the conduct and exploitation of the results of a fault injection campaign that is controlled by means of a database system.

This chapter provides a practical view of the organization and of the features offered by the tool and illustrates its interest thanks to several measures extracted from various experiments. Section 2 describes the overall structure of the tool and how assessment campaigns are conducted. The basic SWIFI techniques used are discussed in Section 3, together with related aspects such as temporal intrusiveness. The notion of activation profile is a major parameter of a campaign and the results analysis. We discuss and present several types of workloads used in our experiments in Section 4. Sample readouts and measures are reported in Section 5. Section 6 concludes the chapter by drawing the main lessons learnt and identifying some perspectives for future development of the tool.

2. OVERALL STRUCTURE OF MAFALDA-RT

The architecture of MAFALDA-RT (Figure 3.3-1) consists of ten *Target machines* (*Intel Pentium*-based PC boards) running the real-time system(s) under test, and a *Host machine* (a *Sun* workstation) that controls and analyzes the fault injection campaigns. The target machines can run either the same system, so as to speed up campaigns, or different systems for comparison purpose. Faults are injected into designated components of the microkernel (e.g., the scheduler, the synchronization manager, the time manager, etc.). The application of a fault corresponds to an experiment. A campaign consists of a series of experiments where fault injection targets the same component of the target microkernel. Each target machine is rebooted between each experiment to purge it from any potential residual errors.

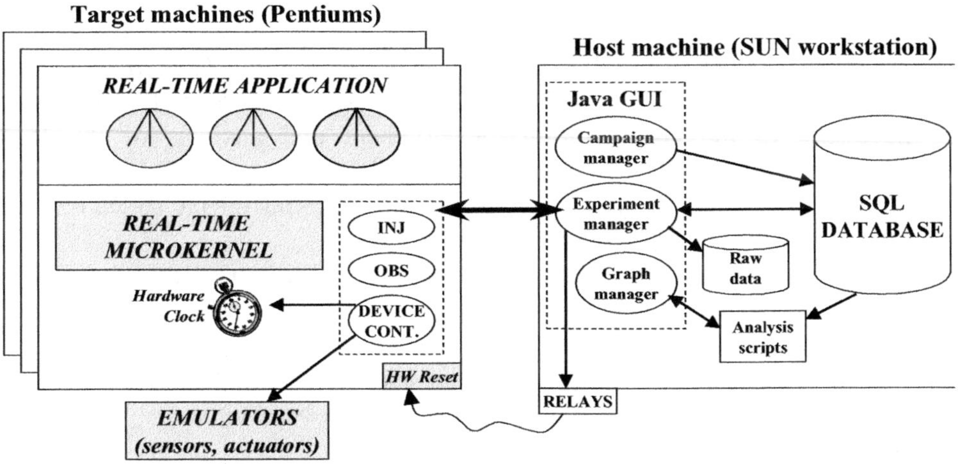

Figure 3.3-1. Architecture of MAFALDA-RT

In the Target machines, MAFALDA-RT comprises the following software modules:

- *Injector (INJ):*
  Consists of a set of exception handlers for fault injection. Faults encompass transient bit flips affecting either the code or data segments of the components forming the microkernel or the parameters of the system calls that are passed to the microkernel. More sophisticated faults (e.g., exchanging the priority of threads) based on the concept of saboteur are also supported (See Section 3.1).
- *Observer (OBS)*:
  Consists of a set of data interceptors monitoring the target system. The interceptors capture temporal and value data that allow for characterizing both its failure modes and temporal performance. Intercepted items include: scheduler events (e.g., context switches), task results, signals (e.g., exceptions), initiation, termination and return code of system calls, etc. All intercepted items are assigned a timestamp.
- *Device controller (DEVICE CONT.)*:
  This module aims at controlling target machine devices so as to eliminate the temporal intrusiveness introduced by the injection and observation instrumentation induced by the SWIFI technique. This is mainly achieved by acting on the hardware clock. External devices, like sensors and actuators, are also controlled by acting on their corresponding emulator software, usually available during the evaluation phase of the system (see Section 3.2).

The core of MAFALDA-RT in the Host machine is the *Database* (a SQL-based database called *PostgreSQL* [POST_03]), which stores all information necessary to the configuration, development and analysis of the fault injection campaigns. The database contains four main types of data, namely, attributes of the target systems, configuration of campaigns, characterization of experiments and experimental results.

MAFALDA-RT provides a Java-based graphical user interface (*Java GUI*) which is connected to the Database. It is composed of the *Campaign manager*, the *Experiment manager* and the *Graph manager*.

The *Campaign manager* allows the user to enter the attributes of the target systems and to configure the fault injection campaigns:

- Attributes of the target systems:
  Characterize the targets of injection. As an example, a target microkernel is described from the point of view of its componentized architecture (e.g., memory addresses of functional components) and its interface (e.g.,

list of kernel calls and parameters). These attributes must be specified for each new target system.

- Configuration of campaigns:
  Consist of parameters defining the fault injection experiments (e.g., fault model, fault trigger, fault duration, etc.), the workload(s) being applied to activate the system (e.g., specific processes, application tasks, etc.), and the supporting hardware platform.

The *Experiment manager* is in charge of controlling the fault injection experiments, but also of storing into files and filtering into the Database the *Raw data* (Readouts) gathered. The filtered data correspond to both the experiment characterization and the experimental results:

- Experiment characterization:
  Informs of the actual activation of the injected faults, the memory addresses corrupted, the time of injection, the fault model, etc.
- Experimental results:
  Correspond to the failure modes (timing and value failures, error detection, etc.) and the performance (processor utilization, response times, laxity, etc.) observed during the fault injection campaigns.

When a Target machine hangs, only a *hardware reset* can reinitialize it. In order to automate a fault injection campaign, a simple relay-based integrated circuit (*RELAYS*) has been implemented to automatically reboot a Target machine at the end of each experiment. This allows for 20 experiments to be run per hours.

Finally, the *Graph manager* allows the user to define under the form of *Analysis scripts* the different types of analyses to be performed on the results stored in the Database, which can then be graphically displayed.

In the subsequent sections we further discuss the capabilities offered by MAFALDA-RT to specify a fault injection campaign according to the fault, activation, readouts and measures attributes.

3. FAULT INJECTION

Among the various forms of fault injection techniques, SWIFI has now become very popular. By flipping bits into processor register cells or in memory, SWIFI allows for emulating the consequences of either transient hardware faults and to provoke more elaborate erroneous behavior that could be induced by software faults [DURA_02]. Thus, SWIFI is a powerful technique that is able to emulate the consequences of faults originating at the various layers of a target system (processor, microkernel, OS, middleware, application software). Many related techniques and tools have been developed to support the application of SWIFI (FERRARI, FINE, Xception, just to name a few).

Nevertheless, although SWIFI exhibits many valuable properties, its application for assessing dependable systems with stringent real-time requirements raises the problem of temporal disturbances that are caused to the target system by the overheads induced by most SWIFI instrumentations currently available. This is also known as *temporal intrusiveness* or *probe effect*. The main sources of temporal intrusiveness are: i) the time related to the injection of faults, and ii) most importantly, the time related to the observation of the system behavior.

After a brief description of the fault models supported by the tool and the means used to implement them, we present the way temporal intrusiveness issues are addressed by MAFALDA-RT.

3.1 Fault models and SWIFI

The first major objective of MAFALDA consists in the analysis of the microkernel behavior in the presence of faults. Such an analysis is intended to evaluate the dependability properties of the microkernel in terms of i) its interface robustness and ii) its internal error detection mechanisms. The former relates to the aptitude of the microkernel to cope with external stress, while the latter concerns the built-in fault tolerance mechanisms of the microkernel.

Consequently, three fault models have been defined:

- corruption of the input parameters of application system-calls (referred to as parameter fault injection)
- corruption of the memory space of the microkernel (microkernel fault injection)
- specific faults affecting the semantic behavior of the microkernel (notion of saboteur).

The first fault model basically simulates the propagation of errors induced by software faults from the application software level to the microkernel level and consists in performing bit flips on system calls input parameters. In practice, this is achieved by targeting the memory image of the *stack* where the parameters of the system calls are passed to the microkernel. The second one is intended to simulate both hardware faults (and to some extent software faults, as well) impacting the code and data segments of the microkernel, by applying bit flips to its memory address space. The third fault model simulates the effect of specific software faults that would affect the runtime behavior of microkernel functions, i.e., thread scheduling, synchronization, etc. The corruption may consist in changing the priority of a thread, giving unrestricted access to a critical section, etc.

Fault injection takes advantage of the componentized architecture of the last generation of microkernels (e.g., *Chorus* [CHSY_96], *LynxOS* [LYNX_00], *VxWorks* [VXWO_98], etc.). Indeed, MAFALDA support the analysis of the impact of the injected faults i) per functional component of the microkernel (e.g., synchronization, scheduling, etc), and ii) between these components (i.e., error propagation). This also leads to the notion of modular workload that aims at activating each microkernel component separately. This is a very useful feature to readily analyze the propagation of errors from one internal component, target of the injection, to another component within the microkernel (See Section 4.1).

3.2 Coping with the temporal intrusiveness of SWIFI

The necessary interactions of SWIFI with the target system can considerably impact the temporal behavior of the target system. Besides such a temporal intrusiveness maybe acceptable in many practical situations, it might not be so when accounting for the stringent temporal requirements (e.g., task deadlines, etc.) attached to a real-time system. Indeed, in such a case, special care is needed so that the potential temporal alterations introduced by the tool (both for fault injection and observation) are minimized or at least mastered.

Towards these ends, the approach supported by MAFALDA-RT consists in providing means for properly controlling the system devices so as to emulate an ideal environment where the temporal overhead of the tool is virtually eliminated. Informally, this approach consists in freezing the progression of time while the tool executes. As a result, the tool execution is perceived as instantaneous by the target real time system.

A real time system is driven by interrupts. Even the notion of time is built from interrupts. Indeed, a hardware clock periodically interrupts the system to inform of the pass of time. Interrupts can be divided into internal interrupts and external interrupts. Internal interrupts are those activated within the target system (e.g., clock interrupts control the release of *periodic* tasks in the real time system). External interrupts are those triggered by the external devices (e.g., sensors); most of the time, they activate *sporadic* and *aperiodic* tasks.

The emulation approach consists thus in properly *stopping* both internal and external *devices* so as to prevent them from triggering interrupts while the tool executes, and then *resuming* them at the end of the tool execution.

The main internal device of a real-time system is the hardware clock. This device periodically interrupts the OS to inform of the pass of time. The hardware clock usually consists of an interval counter that triggers an interrupt whenever it reaches value zero (e.g., the *Intel* 8254 interval timer). Such a device can be usually managed by software so as to disable and resume counting. For instance, the *interrupt on terminal count* mode of the *Intel* 8254 interval timer supports such a feature.

Conversely, when testing a real time system, external devices can be emulated by software. As an example, the behavior of a temperature sensor can be emulated by a program that models a virtual environment where the temperature fluctuates, and that triggers an *interrupt* whenever the temperature increases above a predetermined threshold. When external devices are emulated by software, stopping and resuming their progression consists in stopping and resuming the execution of the emulation program. This whole process is described by the steps shown in Table 3.3-1.

Table 3.3-1. Fault injection steps

1. Stop the hardware clock;
2. Disable the communication port (emulator in a remote machine) or perform an emulation step (emulator in the target machine);
3. Inject a fault and/or observe the system behavior;
4. Enable the communication port (emulator in a remote machine)
5. Resume the hardware clock.

Whenever the whole fault injection process is invoked, the *Device Controller* stops the hardware clock (Figure 3.3-1).

When external devices are emulated by a program that executes on a remote machine, it also disables the communication port connected to this machine, which will halt the execution of the emulator. Conversely, if external devices are emulated in the target machine, the *Device Controller* allows a step of the emulation program to be executed. Note that, in this case, the emulator program runs within the tool context, so its overhead is eliminated. Next, the *Injector* injects a fault and/or the *Observer* monitors the system behavior. Then, the *Device Controller* enables the communication port (if applicable) and resumes the hardware clock. Therefore, the tool is released whenever a fault must be injected or an event must be observed. In addition, if external devices are emulated in the target machine, the tool is also activated whenever a step of the emulation program has to be performed.

Unlike many previous works (e.g., see [DAWS_96] [CUNH_99]), this approach can be used for testing real time systems without having to modify neither the real time application, nor the tool to meet temporal requirements. The main requirement is that an emulation program is to be provided which means that the real physical environment cannot be readily used. Nevertheless, it is worth noting, that emulating external devices by software (e.g., sensors and actuators) is a common practice during the testing phase (e.g., see [KARL_98] [CHEV_01] [HILL_02]).

4. WORKLOAD AND ACTIVATION

The workload applied to activate the system during the fault injection experiments is a very important attribute. Indeed, different activation profiles can lead to distinct behaviors. MAFALDA-RT does not support the specification of the workload to be run. The tool simply offers facilities to select and execute workload processes that are provided as executable files. We discuss some important aspects of this issue hereafter.

Ideally, the activation workload should match the target application profile, so that the results of the analysis can be fully significant with respect to this application. But, in practice, the actual application tasks are not always available. Moreover, when several microkernel candidates are to be compared as part of an early selection process, it is more practical (and most often sufficient) to consider a synthetic and modular workload in which each process is meant to target some specific function of the microkernel, as it is common practice for performance benchmarks.

Accordingly, we have used two types of workloads. In a first step, we use a synthetic workload, composed of a set of workload processes that aim at activating specific microkernel functionalities independently. This kind of workload seldom matches a specific application profile. Nevertheless, it provides a useful activity for selectively testing the various components of a "componentized" microkernel (synchronization, scheduling, etc.). Also, it enables error propagation between such components to be conveniently analyzed. The second kind of workload is an actual real time application for which hard deadlines and other temporal constraints are explicitly specified. This allows for a faithful (albeit, application specific) and comprehensive (encompassing both OS- and application-related properties) robustness testing to be carried out in both the time and the value domains.

4.1 Synthetic workload

The synthetic workload we used is modular and composed of (at least) as many multithreaded processes as functional components of the microkernel. All workload processes run concurrently and perform system calls to each microkernel component independently (e.g., see [ARLA_02]). They are deterministic and thus an Oracle (i.e., a nominal reference) can be provided.

These various workload processes were developed independently of any application domain. They are tailored to exercise almost all services provided by each microkernel component. Because they are separate from each other, workload processes normally fail independently. However, a fault injected in one microkernel component may lead to an application failure in the workload process targeting a different component, which reveals that an error propagation channel exists between the two processes through the microkernel.

Thus, it is worth noting that the global results obtained with these workload processes by no means reflects the behavior of the microkernel in the presence of faults in the case of a significantly different activation profile. Still, such a generic modular workload is of practical interest to discover deficiencies in a target microkernel or identify major differences between alternative candidates. Such a workload is also usually more easily to port to several microkernels than a specific application.

Such a form of synthetic workload is the one that was considered for the analyses reported in Section 5.2.

4.2 Real time application

The use of an application-specific workload can help solve the limitations identified for the synthetic workload, provided that an Oracle can be obtained. Indeed, the real application can be used to analyze the behavior of the actual system, i.e., gathering the microkernel and the application layers, in the presence of faults. This aims at evaluating the overall fault tolerance mechanisms, often application-dependent, that have been developed for a given dependable computer system.

The real-time application we have been using is a mine drainage control system application [BURN_97]. It is intended to control a mining environment. The objective is to pump to the surface mine water collected in a sump at the bottom of a shaft. The main safety requirement is that the pump should not operate when the level of methane gas in the mine reaches a high value due to the risk of explosion. Table 3.3-2 contains the main task attributes of the application and the response times resulting from a fixed priority scheduling analysis, as defined in [BURN_97].

Table 3.3-2. Attributes and response times (times are in milliseconds)

Task	Type	Priority	T	B	C	D	R
CH4 Sensor	Periodic	10	80	3	12	30	25
CO Sensor	Periodic	8	100	3	10	60	47

Task	Type	Priority	T	B	C	D	R
Air-Flow Sensor	Periodic	7	100	3	10	100	57
Water-Flow Sensor	Periodic	9	1,000	3	10	40	35
HLW Handler	Sporadic	6	6,000	3	20	200	79

T: Minimum inter-arrival time (task period) B: Worst-case blocking time
C: Worst-case execution time D: Deadline R: Worst-case response time

The external devices of the application consist of sensors that capture water level, water flow, airflow, and the levels of methane and carbon monoxide. The water level sensor is intended to interrupt the system whenever a high or low water level is detected in the sump, which occurs, at worst, once every 6 seconds. The emulation software for this sensor was thus in charge of releasing task HLW Handler randomly with a minimum interval of 6 seconds between two consecutive releases. Since the other devices are polled by the application, their emulator consisted of a random generator of integers, which provided random values as sensor readings to the tasks.

5. READOUTS AND MEASURES

This section illustrates the type of results that can be obtained with MAFALDA-RT. We first show how the impact of fault injection on a representative real time system can be analyzed. The target system is the *mine drainage control system application* introduced in Section 4.2, that was running on the *Chorus* microkernel (version *ClassiX r3* [CHSY_96]). We then illustrate the insights gained when assessing two different microkernels, namely *Chorus* and *LynxOS (r3.0.1)*.

It is worth noting that MAFALDA-RT systematically checks whether the faulted element is accessed during the experiment, i.e., whether the fault is actually activated. Accordingly, only such cases of activated faults are reported hereafter.

5.1 Assessment of the behavior in presence of faults

In order to illustrate the insights obtained by applying MAFALDA-RT, during the execution of the mine drainage application on the Chorus microkernel, we present the results of three different fault injection campaigns. In two campaigns, the target of the injected faults was the scheduling and the timer components. In both cases, faults were injected by means of *saboteurs*. Two forms of *saboteurs* (*substitution* and *omission*) were used for each component, respectively:

- Substitution: substitute the running task by a lower priority task while the former task is leaving a critical section.
- Omission:
  - omit once the insertion of a randomly selected periodic timer into the timeout queue,
  - omit once the expiration of a randomly selected timer.

These campaigns are thus identified as SCH*ss* and TIM*so*, respectively. A third campaign concerns a specific component — not part of the standard *Chorus* delivery — implementing the so-called Priority Ceiling Protocol that we have developed to provide the basic services (e.g., critical section handling) facilitating the execution of the mine drainage application (campaign PCP*bp*).

Figure 3.3-2 reports the distribution of the *first* fault manifestations observed for the experiments (*#Exp.*) carried out in each campaign. In all three campaigns, for each experiment the interval of observation was about 30 s, after which the target machines were rebooted.

As shown by Figure 3.3-2a, the impact of the *substitution saboteur* was catastrophic: the percentages when the application missed some deadline or delivered incorrect results are respectively 10.7% and 6.5%. Still, most errors were signaled by means of built-in error detection mechanisms: the observed ratios are 27.3% when an alarm detected a deadline miss, and 54.8% when a error status signaled an error in a synchronization system call.

Conversely, 28.7% of the faults injected in campaign TIM*so* (Figure 3.3-2b) led the application to issue incorrect results. Indeed, the error induced by the omission saboteurs prevented periodic tasks CH4 Sensor and CO Sensor from being released, while they are required for the correct computation of results. Nevertheless, 71.2% of the injected faults did not lead to any observable impact. After a careful analysis, we identified at least three main reasons for that: i) internal redundancy of the timeout value of the timers, ii) elimination of alarms while the concerned tasks do not miss their deadline, iii) triggering of alarms that are not related to any application task.

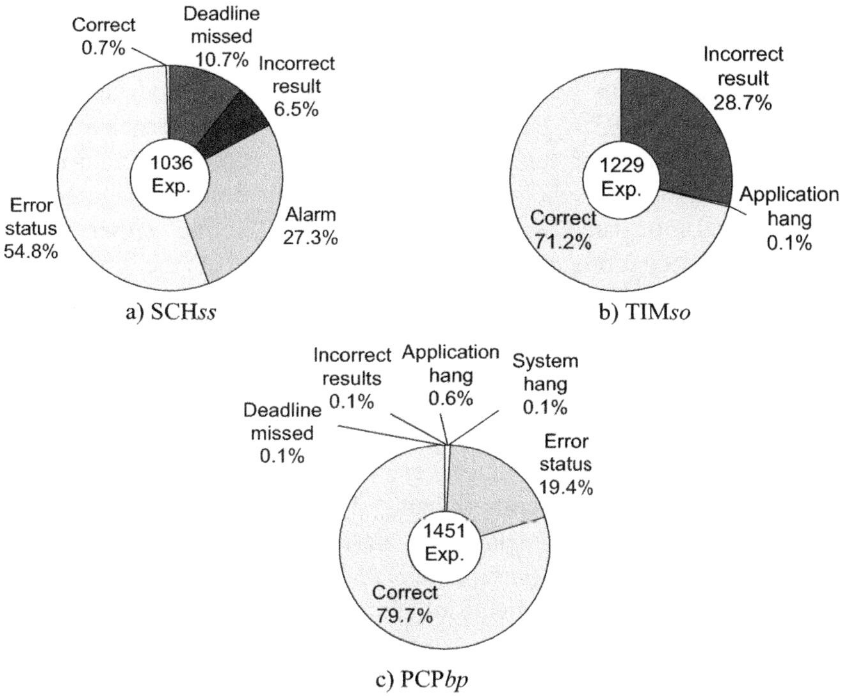

Figure 3.3-2. Saboteurs (substitution and omission) and bit-flips in system call parameters

Few errors impaired the system when the parameters of the synchronization system calls were corrupted (campaign PCP*bp*, Figure 3.3-2c), because of the high percentage of "*No observation*" experiments (79.7%). This is mostly due to the corruption of unused bits within parameters (randomly selection by the fault injection tool). Conversely, the consistency checks implemented within the kernel API (represented by class *Error status*) detected most errors (19.4%). Few errors (0.9%) could thus propagate and lead to the failure of the application.

5.2 Targeting different microkernels

Our objective was not to carry out a detailed comparison of the two microkernels under consideration (*LynxOS* and *Chorus*). Indeed, as already pointed out, many parameters (e.g., target instance) and other aspects of the experimental context (e.g., the activation and fault attributes) impact the results of the experiments. Accordingly, a careful attention must be paid when considering using them for comparison. Our aim was rather to focus on the main distinctive aspects that were exhibited by the campaigns.

The campaigns on both microkernels were conducted on the same grounds (e.g., see [ARLA_02]). In particular, each microkernel was divided into semantically similar functional components, namely: synchronization, memory management, scheduling and communication management. The faults injected consisted in random bit-flips affecting either the code segments of these components or the parameters of the corresponding system calls. A similar synthetic modular workload was used to carry out both campaigns. It is worth pointing out that, while the *Chorus* source code was available to us, this was not the case for *LynxOS*. Besides, the availability of the source code facilitates the definition and execution of a campaign, this could be achieved for *LynxOS* with normal effort.

Figure 3.3-3 illustrates one very distinctive result exhibited by corresponding functional components. The microkernels exhibit very different behaviors when *bit-flips* are targeting the *code* segment of the synchronization component (campaign SYN*bc*). Basically, this component is handling synchronization by means of semaphores or mutexes.

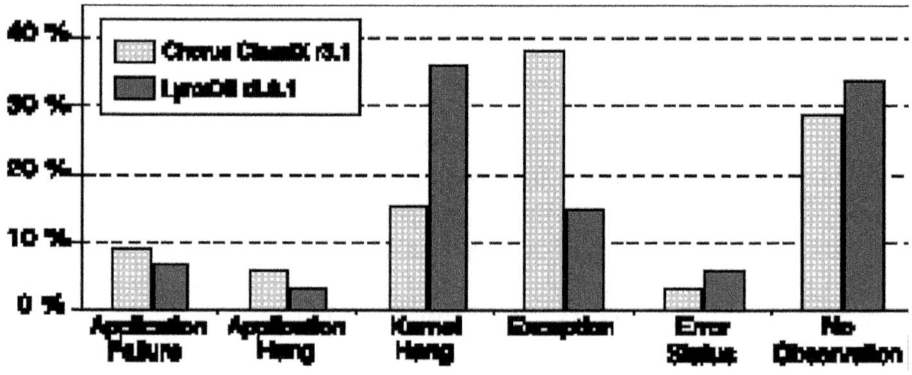

Figure 3.3-3. Campaign SYNbc — Chorus vs. LynxOS

For instance, kernel hangs dominate in the case of *LynxOS*, while for *Chorus* a much larger proportion of exceptions are raised. The results also show that a slightly larger percentage of Error status is reported by *LynxOS*. Such discrepancies reflect the impact of the contrasted design and implementation decisions made by the microkernel developers.

6. LESSONS LEARNT AND PERSPECTIVES

We summarize in this section the various insights we obtained during the various campaigns carried out on the two target microkernels.

Definition of the workload and of an oracle: Clearly, the definition of the workload processes is a major dimension for a system integrator to conduct sensible assessments. They must reflect adequately the use of microkernel services for a given application and an Oracle must be provided.

SWIFI and intrusiveness: Although SWIFI is a flexible fault injection technique that is easy to apply and can simulate a large spectrum of faults, low intrusiveness is another property that is required for fault injection. To minimize intrusion, MAFALDA-RT uses exception handlers separate from the microkernel code to implement the injector and observer modules. Temporal intrusiveness is reduced by the use of the debugging and monitoring facilities offered by modern processors, direct handling of the hardware clock, and emulation of the physical environment. The measurements of failure modes in the time domain (deadline miss, alarm) are thus not disturbed by the injection or the observation mechanisms.

Readout analysis: Beyond global statistical results, the detailed data logged by the tool during the experiments enable the user (microkernel supplier or system integrator) to analyze design/programming flaws. Clearly, the analysis of raw data is required when some singular situation has been observed.

Faulty behavior in the time domain: MAFALDA-RT carries out the analysis of timing properties of the real-time system in presence of faults. In particular, it enables comparing the values experimentally observed for the response time of tasks in each campaign with those analytically predicted by the schedulability analysis.

Interpretation of results: Although depending very much on the activation profile and the configuration of the target instance, useful statistical data can be obtained with MAFALDA-RT thanks to the use of advanced features of a database system. Many SQL scripts have been developed to explore various facets of the logged data. It is open to the user to develop new ones as needed. In particular, a detailed analysis of the ordering of significant events (e.g., detection *vs.* failures) observed during an experiment can be carried out (as shown in [RODR_02_C]).

Integrator vs. supplier viewpoints: From a system integrator viewpoint, the experiments carried out with the MAFALDA-RT help reveal some weaknesses. This is an important input for the selection of a microkernel candidate and also for the definition of error confinement wrappers. From a supplier viewpoint too, the tool can also be seen as a special debugger that provided evidence to be very useful for identifying weaknesses not raised by standard benchmarking, conformance testing and quality assessment.

Porting: The first version of the tool was initially used to evaluate the *Chorus* microkernel. It was then ported to evaluate the *LynxOS* microkernel, in a limited period of time (about 10 man-months). Based on this first experience, we are confident that the porting to a new target should be an easy task, as it involves few modules to be modified (system calls interception, host-target data transfer).

Today, MAFALDA-RT is the most recent and advanced prototype tool of the MAFALDA series. It provides a comprehensive and unique framework that supports both the dependability assessment of COTS microkernels and their integration into computer systems. Although the tool can contribute to sort out various microkernel options available to a system developer, the main practical interest of the tool we would like to emphasize, concerns the design aid capabilities provided to support the integration of the candidate mikrokernel.

The interest of MAFALDA-RT encompasses application domains with stringent dependability requirements (e.g., avionics), but also systems that are critical from an economical viewpoint (e.g., mobile telecommunication) or both (e.g., automobile). The experiments conducted and the results presented illustrate the benefits provided by the tool compared to standard benchmarking and conformance testing procedures.

Future objectives concern application of the tool to other microkernel candidates. Also, we plan to reuse and extend the fault injection and characterization techniques supported by the tool for benchmarking the robustness of other software components i) OSs with respect to faults affecting their drivers and ii) middleware components with respect to OS misbehaviors.

ACKNOWLEDGEMENTS

Frédéric Salles is presently with Sun Microsystems, Palo Alto, CA, USA. The authors wish to thank Arnaud Albinet at LAAS-CNRS and Jean-Michel Sizun, now with Astrium France, for their respective contribution to the development of the user interface and the database of MAFALDA-RT. This work was partially carried out in the framework of the Laboratory for Dependability Engineering (LIS) and supported in part by the EC under Project IST-2000-25425: Dependability Benchmarking (DBench).

PART 4

SIMULATION-BASED FAULT INJECTION

Chapter 4.1

VHDL SIMULATION-BASED FAULT INJECTION TECHNIQUES

Daniel Gil, Juan Carlos Baraza, Joaquín Gracia and Pedro Joaquín Gil

Università Polytecnica de Valencia, Spain

Abstract: This chapter presents an overview of some principal VHDL simulation-based fault injection techniques. Significant designs and tools, as well as their advantages and drawbacks, are shown. Also, VFIT, a VHDL simulation-based fault injection tool developed by the GSTF (Fault Tolerant Systems Group – Polytechnic University of Valencia) to run on a PC platform, is described. Finally, an example of application of VFIT to validate the dependability of a fault-tolerant microcomputer system is shown. We have studied the pathology of the propagated errors, measured their latencies, and calculated both detection and recovery coverages.

Key words: Dependability, Experimental validation, Fault injection techniques, VHDL simulation.

1. INTRODUCTION

In *simulated fault injection*, a model of the system under test which can be developed at different abstraction levels, is simulated in another computer system. Faults are induced altering the logical values of the model elements during the simulation. In the design phase of a system, simulation is an important experimental way to get an early measure of the performance and dependability. Another interesting advantage of this technique respect to other injection techniques is the high observability and controllability of all the modeled components.

A. Benso and P. Prinetto (eds.), Fault Injection Techniques and Tools for Embedded Systems Reliability Evaluation, 41-48.

This chapter is framed in this technique, and particularly in the simulation of models based on VHDL hardware description language. The objective is to show an overview of the main techniques and tools related, illustrating with an application example of the validation of a fault-tolerant system.

The distribution of this paper is as follows. In Section 2 we explain the different VHDL simulation-based fault injection techniques, remarking our implementation contributions. In Section 3 we illustrate the fault models used by the different techniques considered. Section 4 describes the fault injection tool, showing briefly its components and its more relevant features. In Section 5 we present some experiments performed to validate a real fault-tolerant system. Finally, in Section 6 we point out some general conclusions.

2. VHDL SIMULATION-BASED FAULT INJECTION

VHDL (Very High Speed Integrated Circuit Hardware Description Language) has become one of the most suitable hardware description languages from the point of view of fault injection. The reasons for the success of VHDL can be summarized as:

- It promotes an open standard for digital design specification.
- It allows describing a system at different abstraction levels, as it is possible to make behavioral and structural descriptions.
- Some elements of its semantics (such as resolution functions, multivalued types, and configuration mechanisms) can be used in fault injection.

Previous works in this area show that injection in VHDL models can be divided in two groups of techniques, as can be seen in Figure 4.1-1.

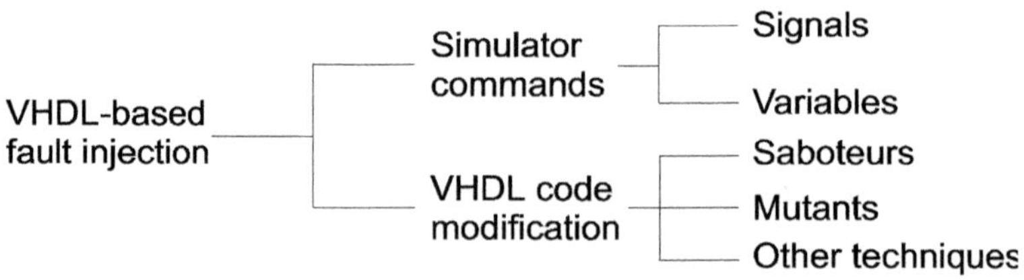

Figure 4.1-1. VHDL simulation-based fault injection techniques

Next we describe the main characteristics of these techniques, emphasizing our own design contributions. We also reference some relevant tools belonging to VHDL simulation-based fault injection techniques.

2.1 Simulator Commands Technique

This technique is based on using at simulation time the commands of the simulator to modify the value of the model signals and variables.

The main reason for using the built-in commands of the simulator for fault injection is that this does not require the modification of the VHDL code; however, the applicability of these techniques depends strongly on the functionalities offered by the command languages of the simulators.

Two techniques based on the use of simulator commands have been identified and experimented: corruption of VHDL *signal* and *variable* constructs that characterize respectively the structural and behavioral features of a VHDL model.

The way that faults are injected depends on the injection place. To inject on *signal*, the sequence of pseudo-commands used is:

1. **Simulate_Until** [injection instant]
2. **Modify_Signal** [signal name] [fault value]
3. **Simulate_For** [fault duration]
4. **Restore_Signal** [signal name]
5. **Simulate_For** [observation time]

This sequence is thought to inject transient faults, which are the most common and difficult to detect. To inject permanent faults, the sequence is the same, but omitting steps 3 and 4. To inject intermittent faults, the sequence consists of repeating steps 1 to 5, with random separation intervals.

The sequence of pseudo-commands used to inject on *variable* is:

1. **Simulate_Until** [injection instant]
2. **Assign_Variable** [variable name] [fault value]
3. **Simulate_For** [observation time]

The operation is similar to the injection on signals, but in this case there is no control of the fault duration. This implies that it is not possible to inject permanent faults on variables using simulator commands.

MEFISTO [JENN_94] was the first injection tool that applied this technique.

It is worth noting that, from the point of view of the injection procedure, VHDL generic constants are managed as "special" variables. This enables injection of some non-usual fault types, such as *delay* faults [GILG_00]. See Section 3 for the description of the fault types that we have applied in this technique.

Respect to implementation cost, simulator commands technique is the easiest one to implement.

2.2 Modifying the VHDL Model

In this category, two techniques can be distinguished. The first one is based on the addition of dedicated fault injection (FI) components acting as *probes* or *saboteurs.* The second one is based on the *mutation* of existing component descriptions in the VHDL model, which generates modified component descriptions called *mutants*. These techniques apply respectively to either structural or behavioral features of the target VHDL model.

2.2.1 *Saboteurs* Technique

A *saboteur* is a special VHDL component added to the original model [AMEN_96_B]. The mission of this component is to alter the value, or timing characteristics, of one or more signals when a fault is injected. The component remains inactive during the normal operation of the system.

Some tools that applied this technique were MEFISTO-L [BOUE_98], [Chapter 4.2] and MEFISTO-C, [Chapter 4.2]. They introduced extensions and improvements over the original tool MEFISTO [JENN_94].

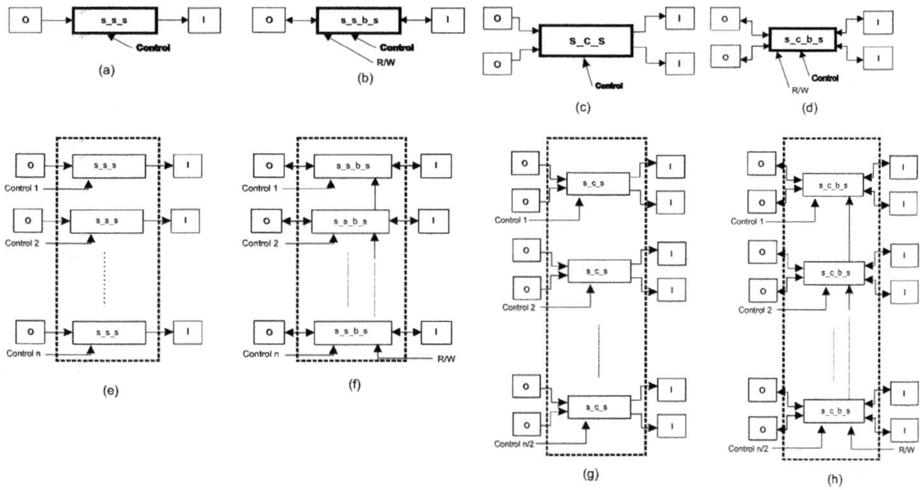

Figure 4.1-2. Different types of saboteurs designed:
(a) Serial Simple Saboteur
(b) Serial Simple Bi-directional Saboteur
(c) Serial Complex Saboteur
(d) Serial Complex Bi-directional Saboteur
(e) n-bit Unidirectional Simple Saboteur
(f) n-bit Bi-directional Simple Saboteur
(g) n-bit Unidirectional Complex Saboteur
(h) n-bit Bi-directional Complex Saboteur

In [JENN_94_B] we can see a classification of *saboteurs*: serial simple, serial complex and parallel. We have extended these types, by designing bi-directional *saboteurs*, and adapting some models to buses [GRAC_01_A] [GRAC_01_B]. With this objective in mind, we have built a *saboteur* library, including the following types of *saboteurs* (Figure 4.1-2):

a) *Serial Simple Saboteur*: It interrupts the connection between an output (driver) and its corresponding input (receiver), modifying the reception value.
b) *Serial Simple Bi-directional Saboteur*: It has two input/output signals, plus a read/write input that determines the perturbation direction.
c) *Serial Complex Saboteur*: It interrupts the connection between two outputs and their corresponding receptors, modifying the reception values.
d) *Serial Complex Bi-directional Saboteur*: It has four input/output signals, plus a read/write input that determines the perturbation direction.
e) *n-bit Unidirectional Simple Saboteur*: It is used in unidirectional buses of *n* bits (address and control). It is composed of *n* serial simple *saboteurs*.

f) *n-bit Bi-directional Simple Saboteur*: It is used in bi-directional buses of *n* bits (data and control). It is composed of *n* bi-directional serial simple *saboteurs*.
g) *n-bit Unidirectional Complex Saboteur*: It is used in unidirectional buses of *n* bits (address and control). It is composed of *n/2* serial complex *saboteurs*.
h) *n-bit Bi-directional Complex Saboteur*: It is used in bi-directional buses of *n* bits (data and control). It is composed of *n/2* bi-directional complex *saboteurs*.

The *control* signals activate the injection. It can be managed by means of the simulator commands, and its activation will determine both the injection instant and fault duration. *R/W* signal determines the data transfer direction. The selection of the fault type can be made using external *selection* signals, managed also by the simulator commands.

The signals where *saboteurs* can be inserted are those which connect components in structural models. The internal architecture of the *saboteurs* can be behavioral or structural. The behavioral design is basically a process whose sensitivity list contains the control and input/output signals. The structural design is based on the use of multiplexers.

The main drawback of *saboteurs* technique is that a number of control signals have to be added to the model. These signals are used to activate one among the set of inserted *saboteurs* and to indicate the type of perturbation that it will have to inject. This adds an additional complexity to both the model and technique. Although *saboteurs* technique is more difficult to implement than simulator commands, it has a larger fault model capability, as it can be seen in Section 3.

2.2.2 Mutants Technique

A *mutant* is a component that replaces another component. While inactive, it works like the original component, but when activated, it behaves like the component in presence of faults.

The mutation can be made in different ways [JENN_94_B]:

- Adding *saboteur(s)* to structural or behavioral descriptions of components.
- Modifying structural descriptions replacing sub-components, i.e. a NAND gate can be replaced by a NOR gate.
- Modifying syntactical structures of the behavioral descriptions.

Given a VHDL model, there could exist lots of possible mutations, but representative subsets of faults at logical level can be considered. [ARMS_92] mentions eight fault models generated by changing syntactical structures of the behavioral descriptions. Table 4.1-1 summarises these eight faults.

Also, in [GHOS_91] several algorithmic models for control flow faults are commented. Some of them coincide with the ones specified in Table 4.1-1. Others include modifications of synchronization and timing clauses (*after* and *wait* clauses).

Figure 4.1-3 shows the method of implementing *mutants* using the configuration mechanism in a multi-level structural architecture [GILB_98].

The mutation affects the components of the structural architecture, where different architectures can be assigned to the same component (entity), but only one is actually selected (in the figure, marked with a thick line). On the other hand, the assigned architectures can be behavioral or structural. In the figure, two configurations are represented: *fault-free* and *mutant*. The "mutated" configurations are obtained by varying the component instances. Depending on the modified syntactical elements, different *mutants* can be obtained from the fault-free original structural architecture. For each component (component i, i = 1... M), it is possible to choose the non faulty architecture or one mutated architecture from the set: (i,2), (i,3),...,(i,N_i).

Table 4.1-1. Fault models by modifying syntactical units in behavioral descriptions

Fault name	Code modification
stuck-then	Replace the condition with true
stuck-else	Replace the condition with false
assignment control	Disturb and assignment operation
dead process	Eliminate the sensitivity list of a process
dead clause	Eliminate a clause in a case
micro-operation	Disturb and operator
local stuck-data	Disturb the value of a variable, constant signal or expression
global stuck-data	Eliminate all value modifications of a variable or signal in an architecture

Notice that to change the configuration, it is enough to modify the architecture-to-component binding in the configuration declaration. After that, the new configuration must be recompiled. It is important to note that this is a reduced compilation, as the different architectures have already been compiled in the library. Every new compilation affects only the configuration declaration, not the *mutant* architectures.

Remark also that the architecture-to-component binding is static. That is, after compiling a specific configuration, it remains unaltered during the simulation. For this reason, only permanent faults can be injected on this way.

The implementation of transient faults by means of *mutants* technique requires the dynamic activation of the mutated configurations. A possible method based on the use of guard expressions in blocks (guarded blocks) can be seen in [BARA_02] [GRAC_01_B]. The idea is that guarded assignments enable blocks and associated architectures to be activated dynamically. By varying the value of the guard signal along the simulation time, it is possible to select dynamically the fault-free architecture or the mutated one.

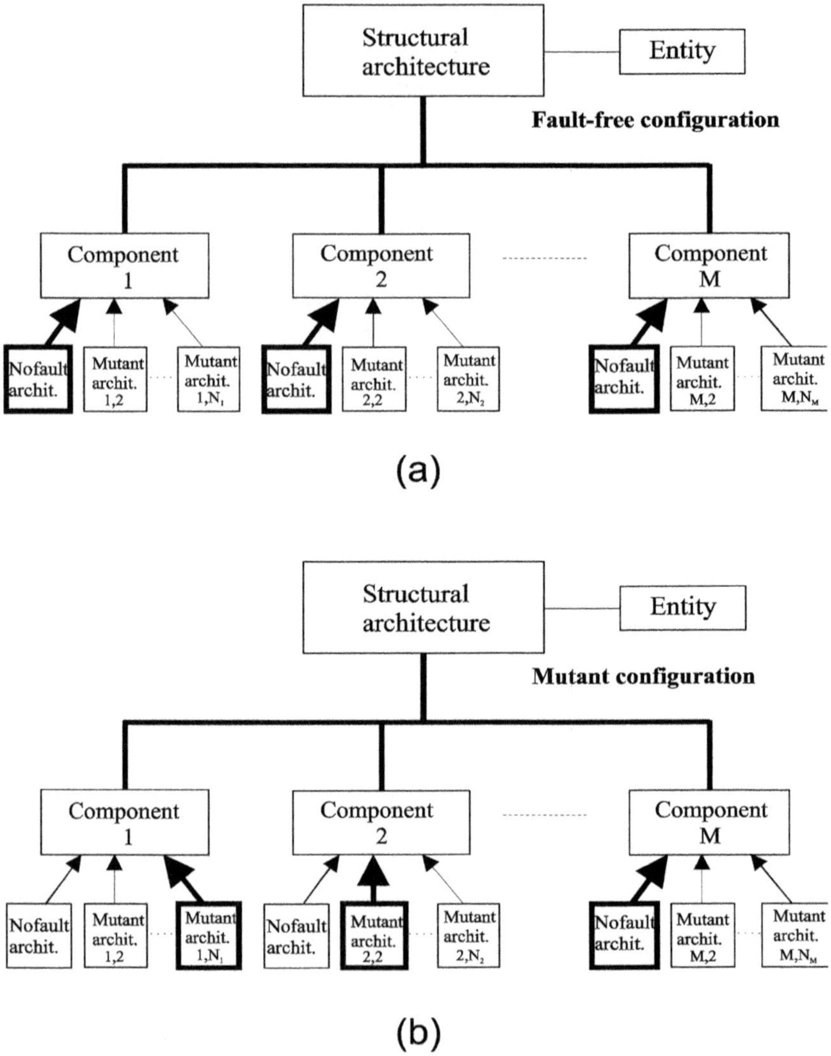

Figure 4.1-3. Implementation of mutants using the configuration mechanism:
(a) Fault-free configuration
(b) Example of mutated configuration

The main problem of transient *mutants* technique is the high temporal cost of simulations, mainly due to the need of saving the system state before architecture commutations. Another general drawback of *mutants* is the need of a high memory space to store all the mutant components. Nevertheless, *mutants* have a big fault modelling capability because they can use all the syntactic and semantic capabilities of the VHDL language. Moreover, fault models are independent of specific IC manufacturing technologies.

2.3 Other Techniques

To avoid the disadvantages of *saboteurs* (additional control signals) and *mutants* (temporal and memory cost), other techniques are implemented extending the VHDL language by the addition of new data types and signals and the modification of the VHDL resolution functions [DELO_96] [SIEH_97]. The new elements defined include fault behavior description. However, these techniques require the introduction of *ad-hoc* compilers and control algorithms to manage the language extensions.

3. FAULT MODELS

An important aspect when working with simulation fault injection is the accuracy of the fault models used. As these models should represent real physical faults that occur in ICs, we have extended the most used models stuck-at (0, 1) (for permanent faults) and bit-flip (for transient faults) with newer and more complex models. This extension is based on new fault models that affect to submicron VLSI technologies. Table 4.1-2 summarizes the fault models that we have applied in every injection technique.

Fault models for simulator commands and *saboteurs* technique have been deduced from the physical causes and mechanisms implied in the occurrence of faults, at device level [CONS_02]. They can be implemented using the VHDL multivalued types std_ulogic and std_logic from the IEEE STD_Logic_1164 package. These packages allow the declaration of signals and variables with a wide range of values in the logic and RTL levels. Std_logic type has also a resolution function to manage the value of output signals connected in parallel.

In relation to the fault models used in *mutants* technique, we have applied changes in syntactical elements of the language, at algorithmic level. As commented in Section 2.3, we have considered the models proposed in [ARMS_92] [GHOS_91]. However, fault modelling for algorithmic behavioral descriptions is an open research field, and new fault models must be studied and tested in the future.

Table 4.1-2. Fault models applied in injection experiments

Injection technique	Transient	Permanent/Intermittent
Simulator commands	Pulse*, Bit-flip**, Indetermination, Delay	Stuck-at (0,1), Indetermination, Open-line, Delay
Saboteurs	Pulse*, Bit-flip**, Indetermination, Delay	Stuck-at (0,1), Indetermination, Open-line, Delay, Short, Bridging, Stuck-open
Mutants	Syntactical changes	Syntactical changes
* In combinational logic ** In storage elements (registers and memory)		

4. DESCRIPTION OF VFIT

4.1 General Features

VFIT (VHDL-based Fault Injection Tool) runs on a PC (or compatible) under Windows™. It has been designed around a commercial simulator (ModelSim, by Model Technology). It is a simple and easy-to-use tool, suitable for injection campaigns into systems of any complexity [BARA_02]. Other significant characteristics are:

- VFIT is model independent, so that it is very easy to apply to any design. The only requirement to be used is a deep knowledge of the model, because to carry out an injection campaign it is necessary to specify a number of parameters that depend on the model.
- The injection tool here presented is fully automated. Once the parameters of the injection campaign have been specified, the tool performs all the phases of the injection (see Section 4.2).
- The intrusiveness of the tool depends on the injection technique used. For simulator commands technique it is negligible, as this technique does not modify the VHDL source code. The other techniques, though, because of their nature alter greatly the original model.
- The performance and limitations of VFIT depend on two factors: a) the computer system where it runs (i.e., on the CPU speed, the amount of RAM memory and hard disk available, etc.), and b) the simulator performance and limitations. The first factor is obvious. About the second factor, the power and flexibility of the command set determine both the simulation speed and the fault models that it will be possible to apply.

- VFIT is able to inject faults into VHDL models at gate, register and chip level. It can apply a wide range of fault models, surpassing the classical models of stuck-at and bit-flip. In fact, it is able to apply all the models indicated in Table 4.1-2.
- Different injection techniques can be used: simulator commands, *saboteurs* and *mutants*. Although the three techniques have been implemented and applied [GRAC_01_A] [GRAC_01_B], only simulator commands technique is fully implemented in the latest release of VFIT [BARA_02].
- With regard to fault timing, permanent, transient and intermittent faults can be injected. To determine both the injection instant and duration, it is possible to choose among different probability distribution functions (Uniform, Exponential, Weibull and Gaussian) verified in real faults.
- VFIT can perform two types of analysis:
  1. Error syndrome analysis, where faults and errors are classified, and their relative incidence and propagation latency are calculated. This kind of analysis is interesting to determine the error detection and recovery mechanisms more suitable to improve the dependability of a system.
  2. FTS validation, where the detection and recovery mechanisms of the FTS are validated. Here, dependability measures are calculated, such as the detection and recovery coverages and latencies.
- Like any other simulation-based injection tool, the main drawback of VFIT is the temporal cost of the injection campaigns. This happens in three cases: i) when the complexity of the model is medium/high, ii) when the workload of the model is complex and requires a long simulation time, or iii) when the number of injections is very high. To face up this severe handicap, VFIT allows to divide the fault injection campaign in *sessions* (consisting in injecting only an interval of the total number of faults), so that each session can be executed in different computers. We must point that, up to date, VFIT does not distribute the sessions automatically. Either, this is the first step to automatically distribute fault injection campaigns in a PC network.

4.2 Injection Phases

An injection campaign consists of injecting a number of faults into the model, according to the value of the injection parameters. For every fault injected, the behavior of the model is analysed, and at the end of the campaign, the value of some specified parameters are obtained. An injection campaign consists of three independent phases:

- **Experiment Set-up**. Here, both the injection parameters and analysis conditions are specified. The most important injection parameters are: Injection technique (simulator commands, *saboteurs* or *mutants*), number of injections, injection targets, fault models for every injection target, fault instant and fault duration distribution, clock cycle duration, system workload and simulation duration. On the other hand, the analysis conditions depend on the objective of the injection campaign (error syndrome or FTS validation).
- **Simulation**. In this phase, two operations are carried out. First, a set of macros is generated: one performs a *golden run* simulation of the model (with no faults injected); the others have the commands needed to inject the number of faults specified in the set-up phase. Then, the macros are executed by the VHDL simulator, obtaining a set of simulation traces: a *golden run*, and *n* with a fault injected (where *n* is the number of faults). This is the most common case when single faults are injected. However, VFIT also can inject multiple faults in a simulation trace.
- **Analysis and Readout**. The *golden run* trace is compared to the *n* fault-injected simulation traces, studying their differences and extracting the analysis parameters of the system.

4.3 Block diagram

Figure 4.1-4 shows the block diagram of VFIT [BARA_00]. From the figure, it can be extracted the main elements of the tool:

- **Tool configuration**. The mission of this module is to configure the tool, considering the simulator a part of it, setting both the tool and simulator parameters.
- **Syntactical and Lexicographical Analyser**. The mission of this module is to scan all the model files, in order to obtain its *Syntactic Tree*. Basically, this tree includes all the possible injection targets of the model. The tree structure reflects the hierarchical architecture of the model, including components, blocks and processes. Depending on the injection technique used, the type of injection targets may vary.
- **VHDL injector library**. This library holds a set of predefined VHDL *saboteur* components that will be included in the model when this technique is used.
- **Graphic interface**. This utility is a wide set of window-based menus, with which the user can specify the injection parameters and analysis conditions needed to perform an injection campaign. With these parameters and conditions the *injection configuration* and the *analysis configuration* files are generated.

- **Injection macro library**. This library is composed of a set of predefined macros used to activate the injection mechanisms according to the injection technique used. The macros are designed from a subset of simulator commands.
- **Injection manager**. This module controls the injection process. Using the injection configuration generated by the graphic interface, it 1) creates a series of injection macros, in order to perform an error-free simulation and the number of fault-injected simulations specified in the parameters, and 2) invokes the simulator to make it run the macros, obtaining the *simulation traces*. An important aspect of VFIT is the fact that the simulator is launched in background, and the duration of every simulation is nearly the same as launched directly from the simulator. That is, the execution time overhead introduced by VFIT is almost negligible.
- VHDL simulator. As indicated before, we have used the commercial VHDL simulator ModelSim, which provides a VHDL environment for IBM-PC (or compatible) under Microsoft Windows™. This is a simple and easy-to-use event-driven simulator. When activated, the simulator executes the file with macros and generates the output trace of the simulation.
- Result analyser. This module takes as input the analysis configuration generated by the graphic interface. According to those parameters, it compares the golden run trace with all the fault-injected simulation ones, looking for any mismatches, and extracting the analysis parameters specified. Depending on the type of analysis, the objective of the comparison is different. Summarizing, the analysis algorithm performs the following actions:
  - In an error syndrome analysis, when a mismatch is found, it is checked if the activated error is effective, testing the workload result. If the result of the workload is incorrect, the injected fault and the effective error are classified. Also, the error propagation latency is measured. Typical results obtained from error syndrome analysis are the percentage of effective errors and their latency in function of error and fault types.
  - In a FTS validation, if no mismatch is found, it considers that the injected fault has produced a non-activated error. If a mismatch is found, it searches for the assessment of the detection clauses to determine whether the activated error has been detected or not. If not, the activated error can be non-effective if the workload result is correct, or produce a failure if the result is incorrect. If the error has been detected, it searches for the assessment of the recovery clauses to determine whether the error has been recovered or not. At the end of

the analysis, the Fault-Tolerance Mechanisms predicate graph (see Figure 4.1-5) is fulfilled [ARLA_93]. This diagram reflects the pathology of faults, that is, the evolution since faults are injected until errors are detected and eventually recovered.

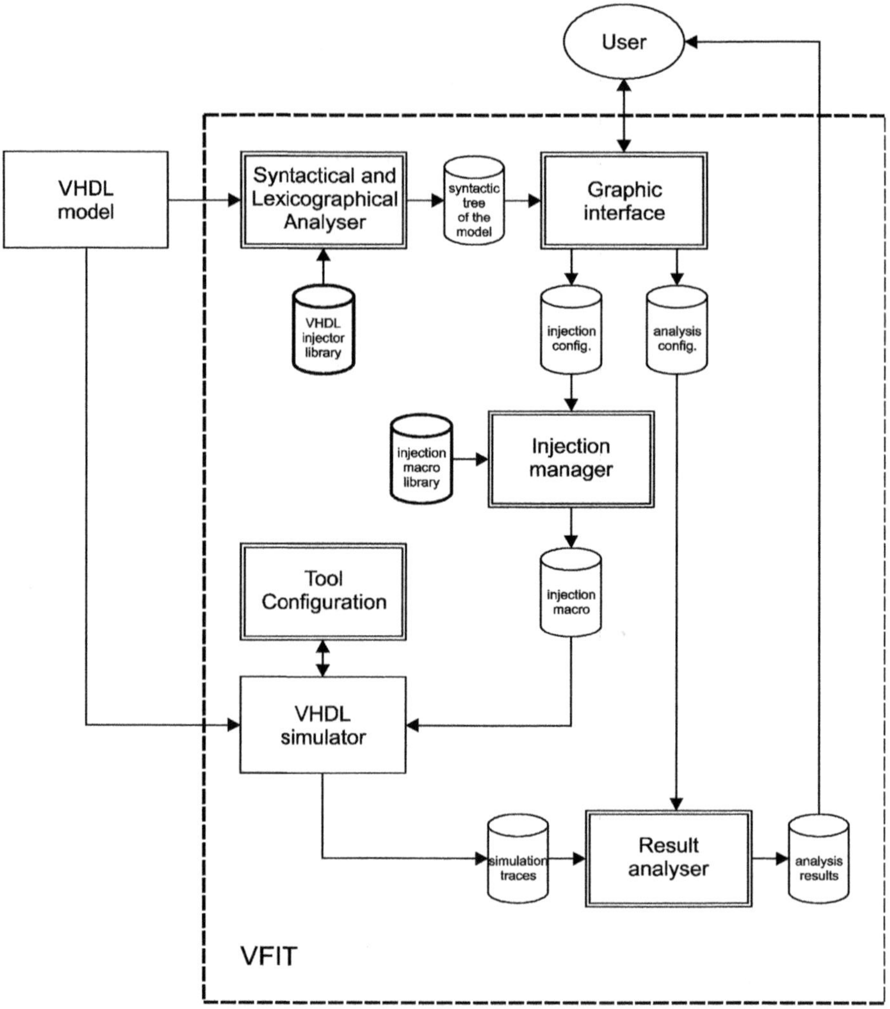

Figure 4.1-4. Block diagram of VFIT

The main measures related to both error syndrome analysis and FTS validation are [GILM_99]: i) Percentage of activated errors, ii) Error detection coverage (detection mechanisms), iii) Error detection coverage (global system), that includes also the number of non-effective errors, due to the intrinsic redundancy of the system, iv) Recovery coverage (recovery mechanisms), v) Error recovery coverage (global system), that include also the number of non-effective errors, and vi) Propagation, detection and recovery latencies.

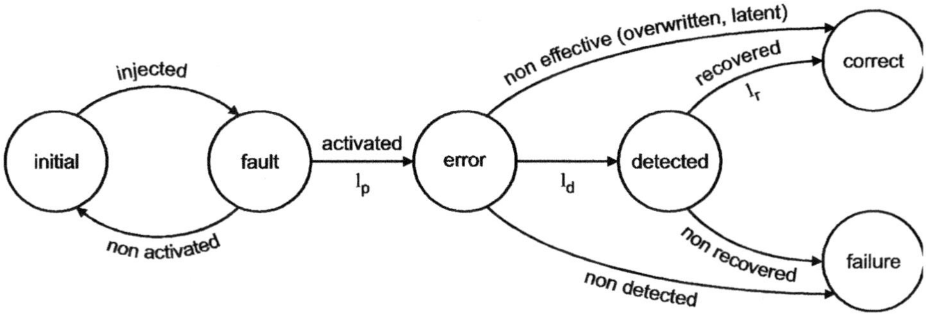

Figure 4.1-5 Fault-Tolerance Mechanisms predicate graph

5. EXPERIMENTS OF FAULT INJECTION: VALIDATION OF A FAULT TOLERANT MICROCOMPUTER SYSTEM

We have built the VHDL model of a fault-tolerant microcomputer. The system is duplex with cold stand-by sparing, parity detection and watchdog timer [GILM_99]. Each component is modelled by a behavioral architecture, usually with one or more concurrent processes. Both main and spare processors are an enhanced version of the MARK2 processor [ARMS_89].

The experiments have been carried out on a PC-compatible Pentium II processor at 350 MHz, and 192 Mb RAM.

Table 4.1-3 shows the conditions of the experiments. The purpose of this injection campaign is to study the response of the system in presence of transient and permanent faults [BARA_00] [BARA_02].

Table 4.1-4 shows the percentage of activated errors, the coverages and the latencies calculated by VFIT. Two types of detection and recovery coverages have been computed: system and mechanisms coverages, as was said in Section 4.3. Finally, Figure 4.1-6 shows the fault tolerant mechanisms graph for two significant cases: short transient and permanent faults.

Independent contribution of the different detection and recovery mechanisms to coverage and latency results can be seen in [BARA_00].

In [GRAC_01_B] the performance of simulator commands, *saboteurs* and *mutants* techniques is compared. Table 4.1-5 shows a comparison of the temporal cost of the *simulation* and *analysis and readout* phases for injection campaigns performed on the same model (running two different workloads: calculus of an arithmetic series and bubblesort algorithm) and computer as in previous example.

Table 4.1-3. Conditions for the fault injection experiments

Type of analysis	Dependability validation
Injection technique	Simulator commands
Number of injections	3000
Injection targets	Any atomic signal and variable of the model
Fault types	Transient and permanent
Fault models	See Table 2
Fault instant	Randomly defined in the range [0-Tworkload], Tworkload = workload duration
Fault duration	1) Randomly defined in the range [0.1T-10.0T], 2) 100 T, 3) Permanent faults
Workload	Calculus of the arithmetic series of n integer numbers

Table 4.1-4. Percentage of activated errors, coverages and latencies, related to fault duration

	Duration		
Dependability measures	[0.1T-10.0T]	100T	Permanent
P_A (%)	23.47	33.40	39.77
$C_{d(mechanisms)}$ (%)	29.40	43.40	47.02
$C_{d(system)}$ (%)	96.31	91.52	88.52
$C_{r\,(mechanisms)}$ (%)	24.29	35.13	18.36
$C_{r\,(system)}$ (%)	91.19	83.23	59.85
L_p (ns)	979	8811	7770
L_d (ns)	31527	45261	40781
L_r (ns)	109915	101879	121464

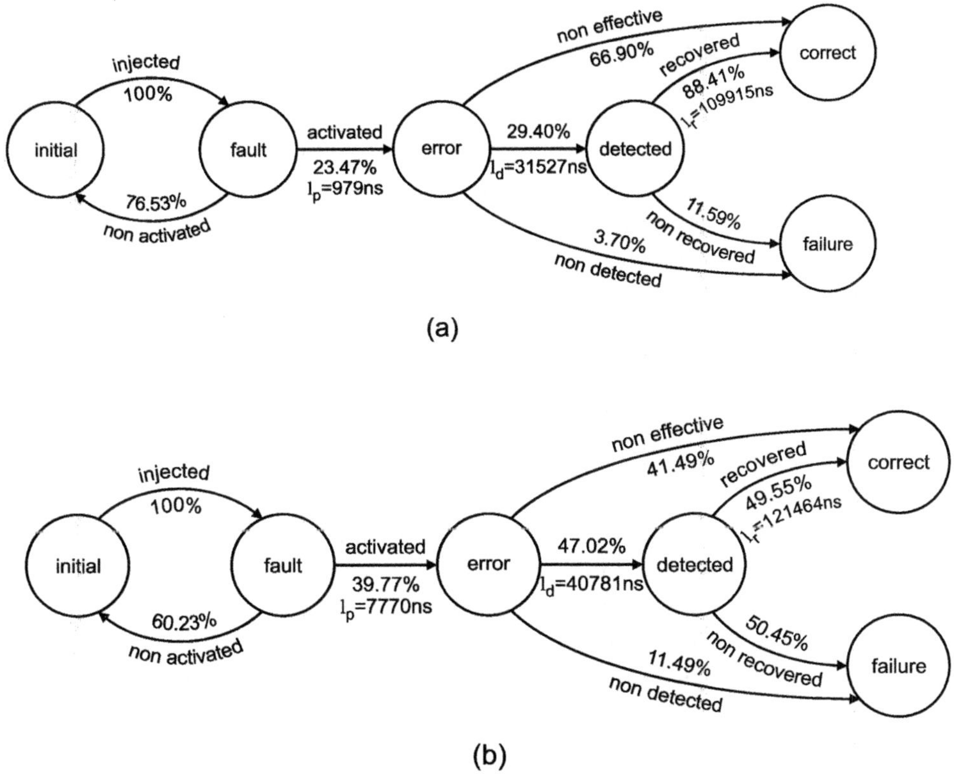

Figure 4.1-6. Fault Tolerant Mechanisms Graph. (a): Transient faults with duration in the range [0.1T-10.0T]. (b): Permanent faults

Table 4.1-5. Temporal cost of VFIT (campaigns of 3000 faults, executed on a PC-compatible PII processor at 350 MHz and with 192 Mb of RAM)

	Simulation phase		Analysis and readout phase	
Injection technique	Series	Bubblesort	Series	Bubblesort
Simulator commands	2 hours	3 hours	2 hours	2.5 hours
Saboteurs	2.5 hours	6 hours	2.5 hours	5 hours
Mutants	10 days	14 days	4 hours	5 hours

Remark in the table the difference in duration of the injection phase between *mutants* and the other two techniques. This is one of most important drawbacks *mutants* technique.

6. CONCLUSIONS

This paper gives an overview of some VHDL simulation-based fault injection techniques. These techniques allow an early validation in the design phase of a system. Another interesting advantage of simulation technique respect to other injection techniques is the high observability and controllability of all the modeled components. The main limitation of simulation-based fault injection is related with the representativeness of fault and system models.

Besides this, VHDL has become one of the most suitable hardware description languages from the point of view of fault injection, because it is a powerful standard for digital design specification. Also it allows describing a system at different abstraction levels, and in addition some elements of its semantics favors fault injection process.

Three main techniques have been shown: simulator commands, *saboteurs* and *mutants*. First one is the easiest to implement, as it doesn't modify the VHDL code. *Saboteurs* and *mutants* techniques extend the fault models, but introduce additional temporal and size costs.

VFIT, a fault injection tool developed by our research group to work in a PC (or compatible) under Windows^{TM} is described, showing briefly its components and its more relevant features.

Finally, it is presented an example of dependability validation of a fault tolerant microcomputer system. We have studied the pathology of the propagated errors, measured their latencies, and calculated both detection and recovery coverages.

Chapter 4.2

MEFISTO: A SERIES OF PROTOTYPE TOOLS FOR FAULT INJECTION INTO VHDL MODELS

Jean Arlat[1], Jérome Boué[1], Yves Crouzet[1], Eric Jenn[1], Joakim Aidemark[2], Peter Folkesson[2], Johan Karlsson[2], Joakim Ohlsson[2] and Marcus Rimén[2]

[1]*LAAS-CNRS Toulouse France*
[2]*Chalmers University of Technology Göteborg Sweden*

Abstract: The early assessment of the fault tolerance mechanisms is an essential task in the design of dependable computing systems. Simulation languages offer the necessary support to carry out such a task. Due to its wide spectrum of application and hierarchical features, VHDL is a powerful simulation language. This chapter summarizes the main results of a pioneering effort aimed at developing and experimenting supporting tools for fault injection into VHDL models. The chapter first identifies the possible means to inject faults into a VHDL model. Then, we describe two prototype tools that were developed using each of the main injection strategies previously identified. Finally, some general insights and perspectives are briefly discussed.

Key words: VHDL, Fault Simulation, Error propagation, Evaluation and Testing of Fault Tolerance Mechanisms

1. INTRODUCTION

The early assessment (encompassing both testing and evaluation) of the fault tolerance mechanisms (FTMs) is an essential task in the design of dependable computing systems. Towards this end, a comprehensive research effort has been carried out with the aim of investigating techniques for fault injection into simulation models of fault-tolerant systems.

A. Benso and P. Prinetto (eds.), Fault Injection Techniques and Tools for Embedded Systems Reliability Evaluation, 41-48.

Due to its wide spectrum of application and hierarchical features, VHDL has been selected as the simulation language to support our work. In addition it is worth pointing out that VHDL offers a viable framework for developing high-level models of digital systems, even before the decision between hardware or software decomposition of the functions takes place.

Elaborating from our pioneering work entitled MEFISTO (*Multi-level Error/Fault Injection Simulation Tool*) that focused on defining and assessing possible means and techniques to inject faults into VHDL models [JENN_94], the research evolved into two specific prototype instances developed and maintained by each institution, namely, MEFISTO-L (at LAAS-CNRS) [BOUE_98] and MEFISTO-C (at Chalmers University of Technology) [FOLK_98]. Besides the tools share a common set of concepts and an overall structure each one features specific fault injection techniques and assessment objectives. In particular, while MEFISTO-L has been targeting the testing of FTMs (for finding potential deficiencies in FTMs, the so-called *ftd-faults*) [ARLA_99], MEFISTO-C has been more concerned with evaluation purpose (e.g., for assessing/comparing the impact of novel run-time injection techniques [FOLK_98] and investigating combined analytical and experimental error coverage estimation [AIDE_02]). These respective characteristics will be briefly described and some lessons learnt discussed.

The chapter is organized as follows.. Sections 2 and 3 describe the main features of MEFISTO-L and MEFISTO-C and also provide examples of their respective application. Eventually, concluding remarks, insights and perspectives are given in Section 4.

2. MEFISTO-L

While the evaluation objective that was the main goal of the original MEFISTO environment can also be supported by MEFISTO-L, the main functional enhancements concern the provision of capabilities specifically aimed at facilitating the removal of *ftd-faults*. The tool was dedicated to supporting the method proposed for fault tolerance testing in the case when the target system and the built-in FTMs are described as a VHDL model. More details on the proposed approach for testing fault tolerance mechanisms can be found in [ARLA_99].

Another major new dimension that has been explored with MEFISTO-L concerns the consideration of techniques for injecting faults consisting in the modification of the VHDL model. The main technique used by MEFISTO-L for injecting faults in a VHDL model is based on the addition of dedicated FI components acting as *probes* or *saboteurs* on VHDL signals. For sake of accuracy and portability, probes are included in the VHDL code only by means of passive derivation, while the insertion technique (simple and bidirectional) is used for saboteurs (Figure 4.2-1).

The other main innovative features of the tool include:

- the embedded VHDL code analyzer, that facilitates the identification of the signals to be injected at different levels of the model hierarchy;
- the observation and injection mechanisms, their synchronization and their automatic placement in the target VHDL model to derive the mutated model (i.e., including *saboteurs* and *probes*).

2.1 Structure of the Tool

MEFISTO-L is composed of three distinct functional blocks: the parsing block, the injection block and the results extraction block (Figure 4.2-1).

The *parsing* block extracts the data needed for the injection campaign from the VHDL source code of the target model. The injection block deals with the whole specification of the campaign, and the generation of the mutated model. The results extraction block analyses the traces obtained from the simulation of the mutated model to produce the campaign results. After the set up of the campaign, the model is mutated to produce the experimental model. *Saboteur* and *probe* components are inserted in the VHDL design units, and the global hierarchy is built. At the top of the model hierarchy, a test bench unit deals with the control of the saboteurs according to the combined values of the probes.

The tool relies on the compilation and simulation capabilities provided by a standard VHDL simulator to compile the mutated model and run the experiments. The tool can be configured with the simulator batch commands to produce compilation and simulation scripts, which can then be launched automatically or manually.

The main objective of the *injection* block is: i) to help the user in specifying the injection campaign and, ii) to automatically generate the mutated model.

The *result extraction* block implements the last step of a fault injection campaign. After the completion of the simulations, the results of the campaign are evaluated, displayed, and saved in a result file.

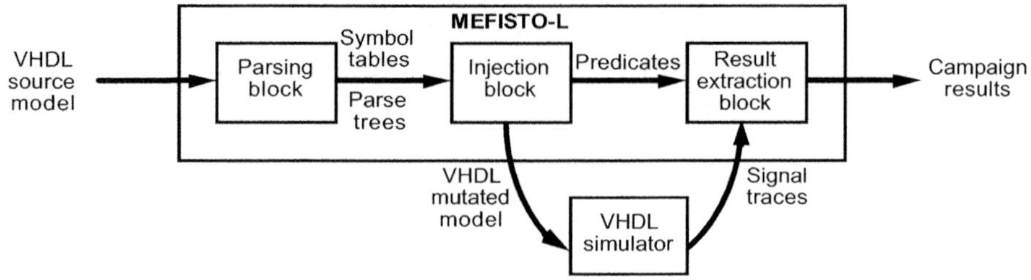

Figure 4.2-1. Structure of MEFISTO-L

Among these various blocks the injection block deserves special attention. Indeed, the testing objective imposes the implementation of sophisticated fault models along with some special mechanisms for observing the system:

- *synchronization probes*, that are meant for synchronizing the injection with a specific activity, especially for testing fault tolerance mechanisms;
- *diagnosis probes*, that observe the behavior of fault tolerance mechanisms through the assertion of pre- and post- predicates (see [ARLA_99]);

The structure of the injection block is described on Figure 4.2-2. To set up the fault injection campaign, the injection block relies on the constructs generated by the parsing block. The main role of the injection block is to support the specification of the injection campaign and, to automatically generate the mutated model.

The main steps encompass the browsing of the source model, the selection and specification of the target and probed signals, the specification of the control of the injection process, and the definition of predicates related to failure mode assumptions. The final step deals with the manipulation of the different parse-trees of the source model mutating the model.

In the sequel, we describe the main features offered by the tool to support these steps for defining the FARM (faults, activation, readouts and measures) attributes that specify a fault injection campaign.

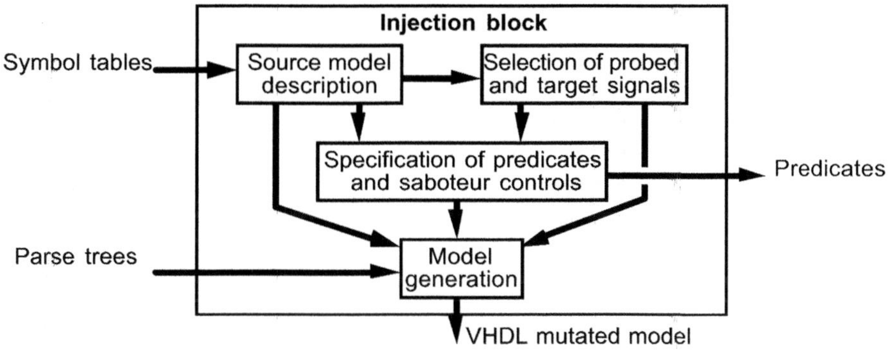

Figure 4.2-2. Structure of the injection block

2.2 The Fault Attribute

The tool gives an overview of the source model to the user, so that the user can easily determine the target signals. Each level of the model hierarchy can be browsed, and described. This allows the user to select directly the target signals in the displayed lists. Figure 4.2-3 illustrates how to select a target signal and apply a fault model on it. The top window is the model hierarchy browsing window. The signal `ext_int_val` is chosen from the input signals of a concurrent signal assignment `node_validation`, and added to the list of target signals for injection. The bottom window gives the list of injectors and probes. In this example, a "stuck-at-0" fault model is applied on the signal chosen in the left side window.

A set of filter mechanisms is available to help build the list according to properties on the signals of the model (declarative regions, names, locations). These filters can be combined with set manipulations functions (union, intersection and difference of signals lists). MEFISTO-L then automatically inserts saboteurs on the target signals. Although a signal consists of an atomic VHDL object, several injection locations may be considered for a signal. Each saboteur has to be assigned with a fault model representing a given disruption.

MEFISTO-L provides a library of saboteurs, including several fault models for main standard types (see Figure 4.2-3). Some fault models are specific to a type, and others are common to several types. The tool allows the user to map a fault model to a group of saboteurs, if the fault model is included in the intersection of the fault model lists for the saboteurs of the group.

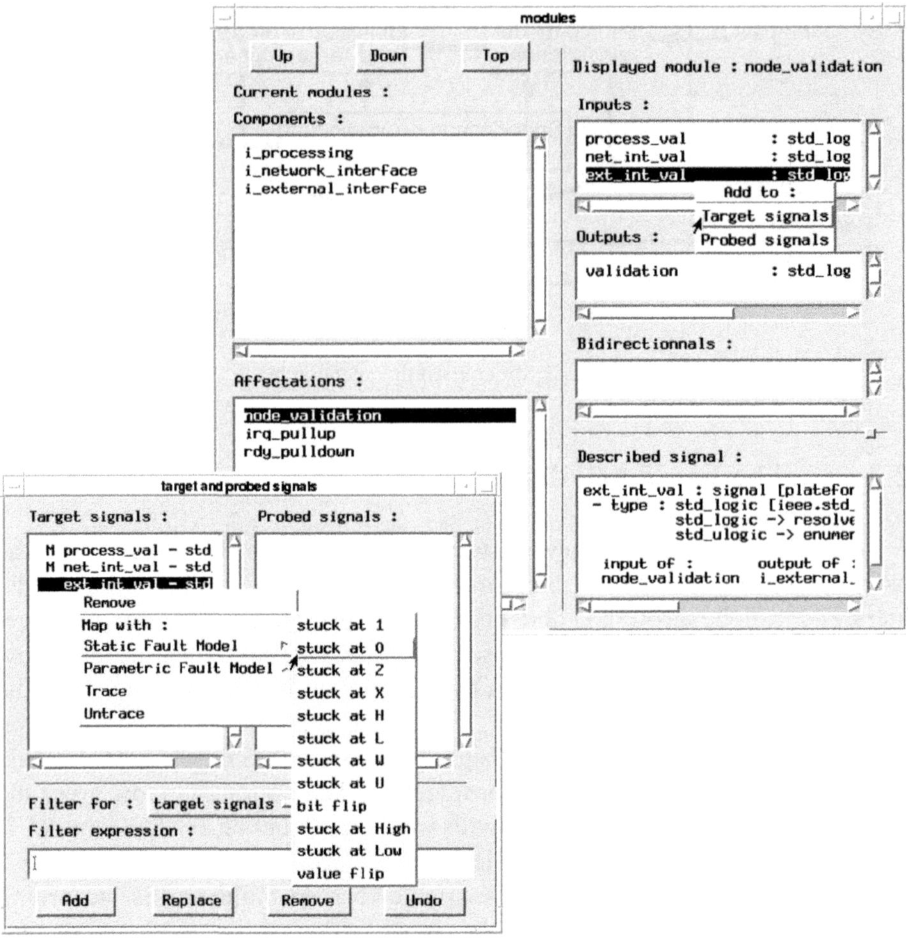

Figure 4.2-3. Selecting a target signal and applying a fault model

2.3 The Activation Attribute

The specification of the fault injection campaign concerns the choice of the fault injection targets, but also the control (activation) of saboteurs. This is achieved by specific mechanisms called probes. As depicted on the top window of Figure 4.2-4, the list of signals to be probed is built at each hierarchical level of the model in the same way as the list of target signals. The probe linked to a signal may be enriched with specifications: these specifications consist in test expressions that will be inserted in the code to communicate internal information to the injection control module of the mutated model. These tests allow dynamic control of the activation of the saboteurs. A probe definition specifies a signal value (or a range of values) for which the probe is asserted as true, and false otherwise. Several probes may be defined for a single signal (indeed a signal might need to be probed both for controlling a saboteur and for being traced as part of the readouts to be collected). In practice, the control of a saboteur is specified at the highest level of the VHDL model hierarchy and relies on a logical expression on some selected parameters from the following list:

- the value of a probe, at every level of the model;
- the result of the comparison between the values of two probed signals;
- the value of static clock probes, defined by the user via their rising edge, their falling edge and their optional period.

These mechanisms make it possible to specify permanent or transient faults, with static or dynamic activation.

The workload executed by the model is another part of the characterization of the activation attribute. The tool does not provide any specific means to specify this workload, besides the definition of file(s) to be executed during the injection campaign. Of course, the type of activity induced by the selected files depends on the goal of the campaign (e.g., testing of fault tolerance mechanisms or coverage evaluation). MEFISTO-L has been used to run workloads mixing control application programs and specific test programs (see [ARLA_99]).

2.4 The Readouts and Measures

Probes and predicates can also be defined for specifying observation conditions to evaluate the outcome of a fault injection experiment. Such predicates are defined as Boolean conditions on the faulted (mutated) model status, or on the comparison of the status obtained during the simulations of the faulted model and of the non-faulted model. In practice, they can be used for handling the testing process as diagnosis pre- and post- assertions / predicates for activating and analyzing specific FTMs (see § 2.1).

Three built-in predicates labeled `error`, `detection` and `failure` are supported by MEFISTO-L with the following measure semantics:

- an error in the faulted model is defined by the simulation instant when the error predicate rises from false to true, and the duration during which it remains true;
- a detection is defined by the simulation instant when the detection predicate rises from false to true;
- a failure is defined by the simulation instant when the failure predicate rises from false to true. The occurrence of a failure stops the simulation process for the current test run, i.e., the values of all the predicates after failure-time are no longer taken into account.

These measures are evaluated at the end of a campaign by the result extraction module (Figure 4.2-1). The traces archived during all experiments are loaded, and the predicates expressions are processed. The error, detection and failure measures are displayed, and saved in a file for further processing.

2.5 Application of MEFISTO-L for Testing FTMs

The applicability and benefits of the strategy aimed at integrating fault tolerance testing activities together with system design activities that has been proposed in have been exemplified using MEFISTO-L.

The proposed strategy features two main activities for the testing of FTMs: i) test pattern generation and ii) test diagnosis. The first activity relies on a statistical testing approach and using the sensitization of the FTMs as the driving criteria to build the test sequence. The definition of pre- and post-assertions/predicates based on the faults affecting the targeted FTMs provides a comprehensive framework to support the second activity. For revealing fault tolerance deficiencies, the testing method aims at generating inputs (errors) of the FTMs which best cover the selected criteria on the design models of the targeted FTMs. To produce these errors, faults are injected in the functional part of the system (together with a precise activation context), rather than in the FTMs.

The overall aim of the proposed validation strategy is to identify relevant testing criteria for activating fault tolerance and adequate readouts for observing the test outcomes, so as to guide the specification of fault injection experiments. These fault injection experiments were then applied with MEFISTO-L to a simulation model of a target fault-tolerant system to adequately exercise the FTMs.

The target system gathers several features extracted from actual fault-tolerant systems for control process in industrial or embedded real-time applications. The architecture features three channels connected by a network supporting a time-sliced protocol. Each channel is designed to be "fail-silent", i.e., in case of detection of an error, the channel should extract itself from the network. The fail-silence property of a channel is achieved by means of a channel extraction mechanism that disconnects its outputs by means of switches controlled by error detection signals.

The VHDL model for this case study represents about 2,500 lines of code, divided in 15 *entities*, 16 *architectures*, 8 *configurations* and 2 *packages* (plus STD and IEEE packages). Table 4.2-1 shows typical times needed by MEFISTO-L to execute this example.

Table 4.2-1. Overhead induced by MEFISTO-L (SparcStation10)

	Design process	FT testing (*n* exp.)
Compilation	70 s	34 s + *n* x 2 s
Simulation	27 s	27 s x n

The first column shows the compilation and simulation times for the VHDL source model. The second column gives the time overhead induced by the FT testing: a) the compilation time is made up of a systematic time + a recurrent time per experiment, and b) the simulation time of the mutated model is not significantly affected. More recently, we have also successfully applied this approach on a more concrete example: the radiation-tolerant processor ERC32 whose VHDL model features 57 *entities* and *architectures*, 3 *configurations* and 19 *packages* in about 26,000 lines of code.

Additional developments still need to be carried out on MEFISTO-L to fully support the proposed testing strategy. They concern in particular the addition of new fault injection capabilities, especially aimed at injecting faults in other VHDL objects than signals: the most promising target is constituted by the variables of *sequential descriptions* in VHDL.

3. MEFISTO-C

The evaluation capabilities of the original MEFISTO tool have been enhanced further in MEFISTO-C. The main focus of the enhancements has been on improving the fault injection efficiency. The time needed to perform fault injection is decreased by the use of a check pointing scheme and by distributing the simulations on a network of slave computers. This section first gives a brief description of the structure of MEFISTO-C and then focuses on the application of the tool.

3.1 Structure of the Tool

MEFISTO-C injects faults in variables and signals defined in VHDL simulation models via built-in simulator commands. The tool offers a variety of predefined fault models to the user. Several options for setting up and automatically conduct fault injection campaigns on a network of UNIX workstations are given.

The operation of the tool resembles the operation of MEFISTO-L. A fault injection campaign is divided into three phases: *set-up*, *simulation* and *data processing* (Figure 4.2-4). The set-up phase involves selecting fault locations, fault models and fault activation times using the MEFISTO-C *Campaign Designer*. In the simulation phase, the MEFISTO-C *Campaign*

Scheduler, *mcs*, executes a VHDL simulator program on one or several workstations. The simulations start with a fault-free simulator run followed by a series of fault injection experiments. The fault-free simulator run is used both for reference and for generating a checkpoint database. The checkpoint database contains the state information needed to start a simulation at a given time. It is used to speed up the fault injection experiments by downloading the nearest checkpoint prior to the time for fault injection to the simulator. Thus, the simulations need not start from the beginning every time. The time needed for the fault injection experiments is made even shorter by dividing the simulations into a short *mandatory* fault simulation part and an optional *extended* fault simulation part. The extended fault simulation part is only executed if the outcome of the experiment cannot be classified after the mandatory part has been executed.

The MEFISTO-C *Campaign Scheduler* is controlled using an X-windows interface. The responsibility of *mcs* is to control the experiment processing in the simulation and data processing phase of a campaign by issuing the tasks of processing experiment batches to slave computers that are ready to take on the issued tasks. Via the X-interface, experiment slaves can be added or deleted. Experiments that are being processed can be stopped, continued or aborted. Additional status information can be obtained about each campaign. This includes information on which experiments have finished and which are left, the time remaining until campaign completion, the rate of completed experiments per hour etc.

In the data processing part, the reference and experiment trace data are analyzed similar to the MEFISTO-L tool to produce dependability measures about the target system.

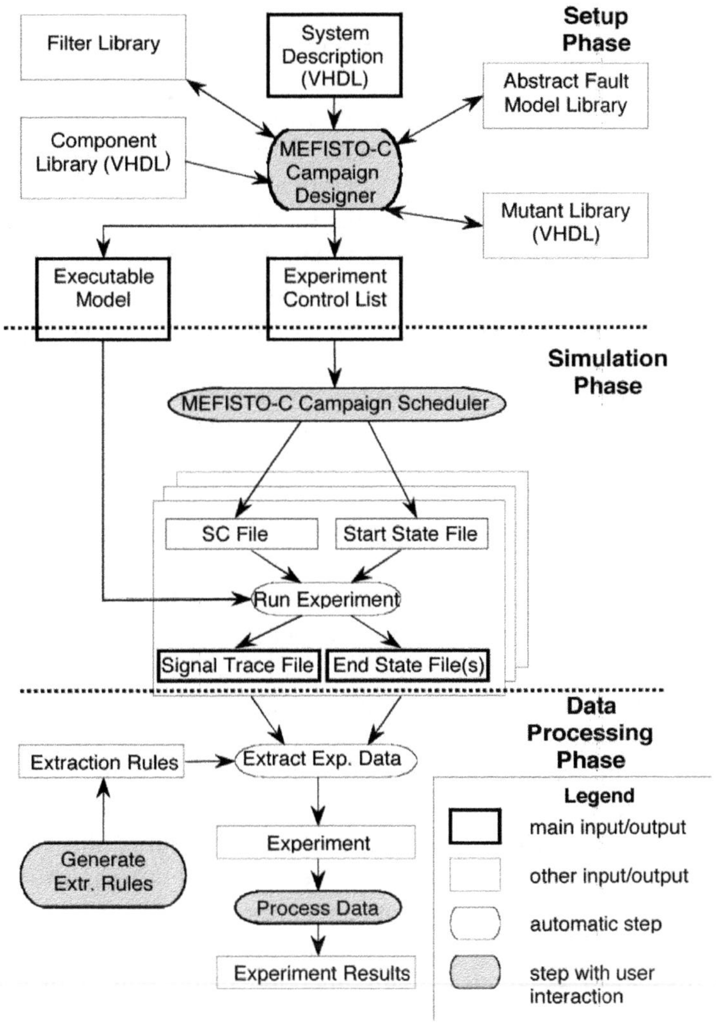

Figure 4.2-4. Overview of MEFISTO-C

3.2 Reducing the Cost of Error Coverage Estimation by Combining Experimental and Analytical Techniques

The *error coverage*, i.e., the probability that an error is detected and handled correctly by the FTMs, is an important parameter for calculating the dependability of a computer system. Estimating the error coverage by fault injection experiments can be very time-consuming because the coverage depends on the activation of the system (cf. Section 3.3). The program executed by the system, as well as the input sequence to the program determines the activation profile.

Thus, to accurately estimate the error coverage, one may need to conduct several fault injection campaigns with different programs and/or input sequences. We have proposed an analytical technique called *Path-Based Error Coverage Prediction* (PBECP, for short) [AIDE_02], aimed at reducing the cost of error coverage estimation. This technique relies on the assumption that we have access to the readouts (cf. Section 3.4) from at least one fault injection campaign where the target system executes a specific program with a given input sequence. The objective is to predict the error coverage obtained when the program receives another input sequence.

We provide here a brief description of the PBECP technique. Readers interested in the details are referred to [AIDE_02]. The basic principle is as follows.

The program executed by the system is divided into basic blocks, i.e., branch free intervals. Depending on the input sequence, each basic block is executed a different number of times. The readouts from fault injection campaigns with one (or a few) input sequence(s) are used to estimate the error coverage for each basic block. The predicted error coverage for another input sequence is then calculated as a weighted sum of the estimated coverage factors for each basic block, where the weight factors are determined by the basic block usage for the input sequence for which the prediction is made.

PBECP can be used in the following way. Assume that we want to identify those input sequences that give extremely high, or low, error coverage among a given set of input sequences. (The set of input sequences of interest is typically determined by studying the usage of the evaluated system.) Instead of conducting fault injection campaigns with each input sequence, we conduct fault injection campaigns with only a few arbitrary input sequences and then apply PBECP to rank all input sequences according to error detection coverage. Once the ranking is done, we can conduct fault injection campaigns with the "interesting" input sequences in order to accurately determine the coverage figures of interest.

This procedure significantly reduces the time needed to identify input sequences that give extremely high or low coverage, as prediction is much faster than conducting fault injection campaigns. The information about the basic block usage for the input sequence for which the prediction is made is collected during a single fault-free simulation of the program execution. The time needed for the prediction is essentially determined by the time it takes to run this simulation. This time is approximately equal to the time it takes to make one fault injection experiment, i.e., to observe the effect of a single fault.

The prediction technique has been used for estimating the coverage of the hardware implemented runtime checks included in the *Thor* microprocessor (a processor designed for use in highly dependable space applications). MEFISTO-C was used to inject single bit-flips inside a VHDL model of the microprocessor to emulate the effects of *Single Event Upsets* (SEUs) caused by ionizing particles in the space environment. The results showed that the estimated error coverage for various input sequences to a *Quicksort* program executing on *Thor* may vary between 92% and 96%.

A comparison of predicted and observed results using three fault injection campaigns conducted with different input sequences to the *Quicksort* program showed that the technique is able to identify the input sequence with the highest or lowest error coverage provided that the difference in actual coverage is significant.

The experiments also indicated the time savings to be expected from using PBECP. The time needed to estimate the error coverage for *Quicksort* with a particular input sequence using fault injection (3,000 experiments) on a 300 Mhz *Sun* workstation is in the order of one hundred hours, while the time needed to predict the error coverage is in the order of minutes.

Although this study has showed promising results, it should be stressed that this research still is in an early stage. More fault injection experiments are needed to show if the technique can be generally applied on other workloads and target systems.

3.3 Using MEFISTO-C for Assessing Scan-Chain Implemented Fault Injection

The feasibility of using MEFISTO-C as an aid in the assessment of another fault injection technique, called scan-chain implemented fault injection (SCIFI), has been demonstrated [FOLK_98]. SCIFI uses the built-in test logic available in many modern integrated circuits for injecting faults. The IEEE 1149.1 standard, commonly known as JTAG since it was developed under the sponsorship of the *Test Technology Committee* of the IEEE and the *Joint Test Action Group* (JTAG), describes the boundary scan test logic. The test logic is commonly used for testing everything from integrated circuits and printed circuit boards to whole systems. The technique is to form a scan-chain by shifting bits into a register through a serial input link enabling the logical values of the pins or internal locations of the circuit to be set. At the same time, the old values of the pins or internal locations are shifted out through a serial output link allowing the values to be read. Faults are injected via the scan-chains by shifting out values through the serial link (requiring dummy values to be shifted in), changing one or several of the values shifted out and then writing back the fault injected values.

The IEEE 1149.1 standard only defines the use of a *boundary* scan-chain, allowing the values of the boundary elements (pins) of the circuit to be accessed. However, other types of scan-chains are common. An *internal* scan-chain can be used to access internal state elements which may be part of both programmer-accessible and hidden registers. A *debug* scan-chain can be a valuable aid for software debugging in a microprocessor allowing break-points to be set, i.e., the execution to be halted, according to various trigger conditions (address reached, data accessed, time-out etc.). The type of scan-chain to access via the serial link is selected by shifting bits into an instruction register. A small portion of each scan-chain may also be used for controlling which parts of the circuit to be accessed. Programming the scan-chain by setting the bits in this portion allows a scan-chain of limited size to be used to access a large number of locations.

Fault injection experiments have been performed on the *Thor* microprocessor for evaluating its capability of handling SEUs. To model the effects of such faults, single bit-flips were injected into the internal state elements of *Thor*. The results from experiments with MEFISTO-C on a highly detailed VHDL model of *Thor* were compared with faults injected via an internal scan chain using SCIFI. By investigating the differences in the results obtained for the two fault injection techniques, the accuracy of the SCIFI technique could be assessed.

The key characteristics of the two techniques applied on *Thor* are summarized in Table 4.2-2.

Table 4.2-2. Comparison of fault injection with SCIFI and MEFISTO-C on the Thor CPU

	SCIFI	**MEFISTO-C**
No. of fault injection locations (state elements)	2,250	3,971
Time resolution for fault injection	one instruction	one clock cycle or less
No. of observable state elements	2,913	3,971
Efficiency (faults injected per hour)	~2,000 faults/h	~7 faults/h per workstation used

Figure 4.2-5 shows a comparison of the results obtained using SCIFI and MEFISTO-C on *Thor* for two different workloads, a digital state model control application and a program implementing a *Quicksort* algorithm. The diagram shows the percentage of errors, which were either latent, overwritten, detected by the various error detection mechanisms of *Thor*, or lead to undetected wrong results. These results are based on roughly 10000 injected faults per technique and workload. Only minor differences in the

distribution of detected errors were obtained for the two techniques while the differences between the non-effective errors (latent or overwritten errors) obtained for each technique are somewhat greater, mainly due to a lack of observability for the SCIFI technique. The percentage of non-effective errors is also lower for the SCIFI technique suggesting that the smaller number of fault injection locations available for the SCIFI technique (2,250 locations *vs.* 3,971 available in the simulations) is particularly sensitive to faults. This conjecture is supported by the fact that the overall error coverage, i.e., the percentage of errors which were either latent, overwritten or detected by an error detection mechanism, is lower for the SCIFI technique than for simulation based fault injection (90%-94% using SCIFI compared to 94%-96% using MEFISTO-C).

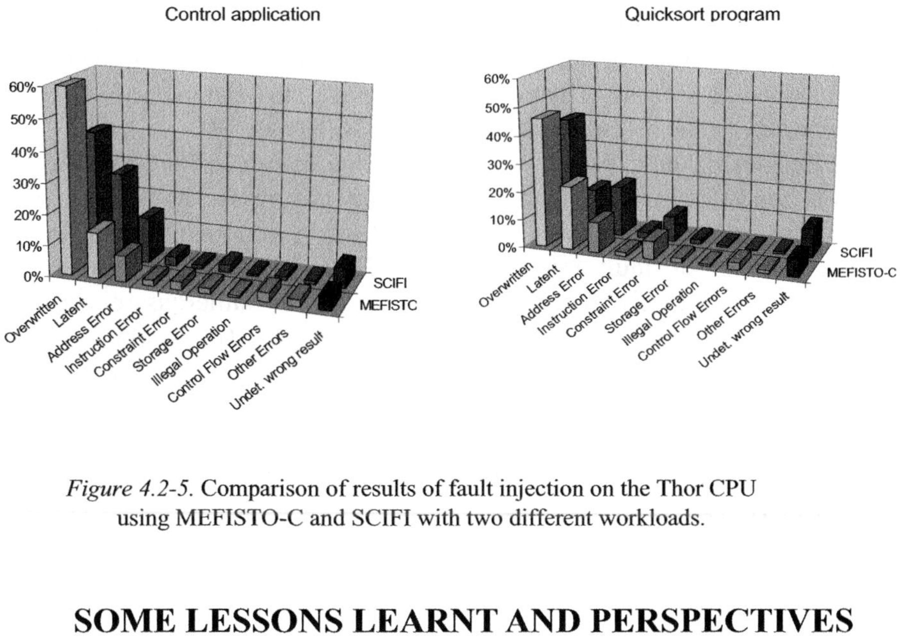

Figure 4.2-5. Comparison of results of fault injection on the Thor CPU using MEFISTO-C and SCIFI with two different workloads.

4. SOME LESSONS LEARNT AND PERSPECTIVES

The wide variety of usages of fault injection into VHDL models that were carried with the prototypes tool reported showed both the feasibility and suitability of the fault injection techniques proposed. The large variety of experiments and studies conducted resulted in broad set of insights relevant to the design and analysis of dependable systems, encompassing: detailed analysis of the fault activation and error propagation processes, the validation of error models, the testing of fault tolerance mechanisms, the assessment of fault injection techniques (SWIFI, SCIFI).

Besides the original, yet classical complete description of a non-pipelined 32-bit microprocessor by Peter J. Ashenden in 1990 and the benchmarks circuits described in [CORN_97], the increased availability of VHDL models for real designs (e.g., ERC-32 and LEON processors supported by ESA-ESTEC; a model for the fault-tolerant version of the LEON processor [GAIS_02] can also be obtained) testify of the timeliness of the techniques as those supported by the series of MEFISTO prototype tools and other related tools.

VHDL has maturated into a recent VHDL-2001 standard as reflected in [ASHE_01] and is now widely recognized as a modeling and simulation language for various stages of the development process of digital systems and is being used in a wide segment of industry (e.g., the *SAAB Ericsson Space Thor* processor targeted by MEFISTO-C, etc.). Accordingly, the testing and evaluation strategies as well as the injection techniques described herein have now become very much suited to be integrated into the development process of dependable systems. This statement is in particular supported by the recent work reported in [BERR_02_B].

Besides of the capabilities offered by the features offered by the pioneer tools reported in this chapter, significant progresses have been made to develop improved methods to support fault injection into VHDL models. The domain of fault injection into VHDL models has now become a flourishing research area, as exemplified by the numerous related projects and works being conducted (e.g., see [GRAC_01_B]). Among these recent and on-going studies, the following specific features are also worth mentioning:

- The explicit account for the fault occurrence rate in the selection of the Fault attribute allows for dependability measures to be estimated [SIEH_97].
- The technique described in [DELO_96] allows the user to apply fault injection at the Instruction Set Architecture (ISA) level, where a behavioral model of a processor written in VHDL is able to execute actual machine code.
- Significant speed-up in the execution of fault injection campaigns run on a VHDL simulator can be achieved by the method proposed in [PARR_00];
- The fault collapsing rules depicted in [BERR_02_B] to reduce the size of the faults to be simulated that rely on the inspiring work reported in [BENS_98_B];
- The VHDL-based statistical testing strategy of FTMs to which MEFISTO-L contributed could benefit from the language extensions and global verification environment reported in [CAMU_94].

ACKNOWLEDGEMENTS

This work was supported in part by ESPRIT Project no 6362 PDCS2 (Predictably Dependable Computer Systems). The LAAS team wants to acknowledge the contributions of Jean-Etienne Doucet and Philippe Pétillon (who is now with *AirBus France*, Toulouse) to the development of MEFISTO-L. The Chalmers team would like to thank Professor Emeritus Jan Torin, the former head of the Laboratory for Dependable Computing, for his long lasting support and encouragement. We also acknowledge the support from *Saab Ericsson Space AB*, that provided the VHDL model of the Thor microprocessor. J. Boué is currently with *Silogic* and E. Jenn with *Thales Avionics* in Toulouse. J. Ohlsson and M. Rimén are respectively with *Carlstedt Research and Technology AB* and *Semcon AB* in Göteborg.

Chapter 4.3

SIMULATION-BASED FAULT INJECTION AND TESTING UNSING THE MUTATION TECHNIQUE

Chantal Robach and Mathieu Scholive

LCIS-ESISAR, Valence, France

Abstract: In contrast to traditional generation testing methods, the proposed approach considers faults that may occur at the functional level such as software faults. This Chapter shows therefore that mutation testing, up till now only used in software, is also highly efficient for hardware testing. At the functional level, mutation testing is a powerful and systematic method to detect design faults. Indeed, it guarantees many standard criteria such as instruction, branch, predicate and boundary value coverage. At the gate level, it was shown that mutation testing (carefully adapted to hardware) also detects efficiently hardware faults. To prove this method experimentally therefore, we have created a tool, named ALIEN, which has been implemented for the VHDL language.

Key words: Functional testing, Mutation test, design validation, VHDL

1. FAULT INJECTION TECHNIQUE: MUTATION TESTING

1.1 Introduction

The increasing complexity of integrated circuits in terms of the number of gates to be processed by gate-level ATPGs and the use of hardware description languages for specifying these circuits at a high level of

A. Benso and P. Prinetto (eds.), Fault Injection Techniques and Tools for Embedded Systems Reliability Evaluation, 41-48.

abstraction, have lead to the introduction of new high level test generation methods. These allow the testing process to be shifted up to the specification level. It has previously been shown [ALHA_99] that a test method, the mutation-based testing method which was originally developed for software could well test a piece of hardware given by its VHDL description, in other words by a program. At the functional level, mutation testing guarantees a set of standard software test metrics which can be used as design validation metrics. At the gate level, the validation functional data are enhanced so that classical hardware faults are also detected.

The mutation-based testing was originally proposed by Budd in [BUDD_78] with the aim of appraising the efficiency of a set of test data for revealing faults present in a given program. The method assumes that a fault is a single syntactic modification in the "correct" program. In the process of mutation testing, the faults are injected in the original program to create "erroneous" versions which are called *mutants* and which can be distinguished from the original program by a set of test data.

The mutation-based testing has been essentially used in the filed of software tests. The main study is the one from the Georgia Institute of Technology [KING_93] which developed a complete test environment: the Mothra Software Test Environment. This tool defines the data which can detect the presence of specific faults during unit or module testing, for the FORTRAN language [KING_91]. Their strategy is based upon the representation of the conditions under which the mutants will be killed as simple algebraic constraints. This approach was called "constraint-based testing" [DEMI_91]. Another use of the mutation-based process is the software-implemented fault injection for fault-tolerance purposes [JENN_94]. This type of approach was motivated by the difficulties and the cost of physical fault injection, and the ability of such a method to discover errors generated both by software and hardware faults. The developed system, called MEFISTO, is an integrated environment for applying fault injection into simulation model and more precisely modification of the VHDL model. The MEFISTO environment is presented in detail in Chapter 4.2.

1.2 Mutation Testing

Firstly, note that the mutation testing is based on the three hypotheses:

- The *competent programmer hypothesis*: a competent programmer tends to write programs that are close to being correct.

- The *coupling effect*: a set of test data that distinguishes all programs with simple faults is sensitive in that it will also distinguish programs with more complex faults.
- The *mutation operators*: an adequate set of predefined operators is used to model all simple faults for a program (the mutation operators are defined according to the programming language).

From these hypotheses, we add the *oracle hypothesis* - the means to verify whether the program's result is correct or not for a given test data - that is used for all software testing techniques.

After the general view on testing mutation, we are interested more precisely in the mutant generation problem. In fact, it is not reasonable to generate mutants so that each mutant corresponds to a possible fault that the programmer can commit [BUDD_81]. The mutation operators are chosen as "sensibility indicators" for the sets of test data to the small modifications in the program under test. The mutation method is even more justified for languages in which small syntactical modifications also produce small semantic modifications. In reality, it is more difficult to detect a small semantic modification than to detect a big one. A set of test data, which can detect small modifications, should be capable of detecting big ones. It is the coupling fault principle that ensures this property, provided that the mutation operators are well chosen.

The big problem when using the mutation testing is to define a minimal and adequate set of mutation operators. After determining the set of mutation operators, to show that it is significant for software testing, it is necessary to demonstrate that a set of test data, detecting the generated mutants from this set of mutation operators, also satisfies some usual criteria like statement coverage, branch coverage, condition coverage and extreme value testing. The mutation operators inject small semantic modifications in the program such as the replacement of an addition by a subtraction, the replacement of a variable by another one of the same type, etc.

Note that a mutant is obtained by injecting only a single modification on original program. Figure 4.3-1 shows an example: the MAX function (written in the PASCAL language) which returns the maximal value of two integers, with four overlapped mutants. Each mutant is represented only by the mutated statement that is labelled by the Δ symbol and is introduced after the original statement. For the first and second mutants, the relational operator is replaced by another one of the same family. The third mutant, the ABS, replaces a variable by its absolute value. The last mutant replaces a variable by another variable of a compatible type.

```
1 function max (m, n : integer) : integer ;
2 begin
```

```
    if (n > m)
Δ   if (n < m)
Δ   if (n ≥ m)
    then Max := n
Δ   then Max := ABS(n)
    else Max := m
Δ   else Max := n
  end (* Max *);
```

Figure 4.3-1. The Max function with four mutants.

Once all the mutants are generated for a program P, the mutants of P and P itself are executed on a set of test data T. If a mutant gives a result different from the one issued by P, the mutant is *detected.* In this case, this mutant is considered as *killed* and then eliminated from the test process. Otherwise, the mutant is considered as alive. Figure 4.3-2 shows the general principle of the mutation testing.

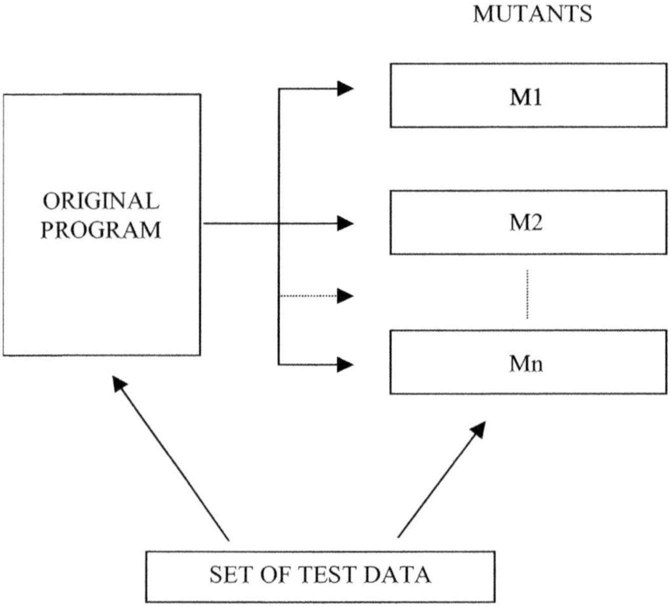

Figure 4.3-2. General principle of the mutation testing

Once all test vectors are executed, each live mutant belongs to one of two cases: or the mutant is functionally equivalent to the original program, either the mutant is detectable but the set of test data is not adequate to kill it. In the first case, no test can kill such mutants, called equivalent (between 4% and 10% of generated mutants [DEMI_91]). It is very difficult to know whether the mutant is equivalent or the set of test data is insufficient to detect it. In practice, the programmer is responsible for detecting the equivalent mutants. In the second case, the set of test data is not powerful enough to kill such mutants. More new test vectors then need to be generated to strengthen the set of test data and the process must be repeated until the tester is satisfied.

Moreover, the mutation testing associates a metric with the set of test data to estimate how well it carries out the program. This metric is expressed by the following mutation score: $MS(P, T) = K / (M - E)$ with K the number of killed mutants, M the number of generated mutants and E the number of equivalent mutants, for a program P and a given set of test data T.

This metric, the mutation score, has an advantage over the other software metrics in that, even in the case where no fault was detected, we can affirm that certain faults cannot exist in the program. One says that a test, which gives a mutation score of 100% is relative adequate with regard to the mutation operators. Finally, as this metric increases our confidence in the program correction, it can be considered a metric of test quality.

1.3 Different mutations

We have just presented the different characteristics of the mutation testing, that is also called *strong mutation testing*. We shall now describe some different mutation techniques that already exist.

1.3.1 Weak mutation

Weak mutation, proposed by Howden [HOWD_82], does not consider either the competent program hypothesis or the coupling effect hypothesis. This technique is described as follows: suppose a program P and a component C of P, C' is a mutated version of C, and P' containing C' is a mutant of P. The test t is constructed so that C and C' compute at least a different value when executing t by P and P', however P and P' can produce the same result.

The other differences between strong mutation and weak mutation are:

- In strong mutation, the functions are considered on the program level, whereas in weak mutation they are considered on the statement level and expression.
- The mutation operators in strong mutation depend on the programming language used, whereas in weak mutation the mutation operators are generic and they do not depend on language.
- There does not exist a global method to generate test data that reveals faults predefined by the mutation operators. In fact, the methods based on

this approach should ensure that the mutant has a behaviour different from that of the original program and this is at program level. It is a very difficult task to generalize. On the other hand, the methods based on weak mutation should ensure a different behaviour only at statement level. This makes possible the establishment of generalized methods for test generation, but to the detriment of a weaker set of test data. It is the reason for the term of weak mutation.

The five types of component considered in weak mutation are:

- Variable reference: replacing one variable reference by another.
- Variable assignment: assigning the value to another variable.
- Arithmetic expression: the expression is modified by an additive constant, a multiplicative constant or a wrong coefficient.
- Arithmetic relation: the expression is modified by a wrong relational operator or an additive constant.
- Boolean expression: the expression is modified by using the preceding operators for sub-expressions.

Weak mutation testing has some advantages over strong mutation testing. It is not necessary to execute mutants in order to choose test data. The number of test vectors is often small for most mutations. A big advantage is also that it is possible to specify a priori the necessary test data so that the mutation gives an incorrect result. Weak mutation does not, however, ensure that the chosen test data for detecting a mutation gives an incorrect result for the whole program.

1.3.2 Firm mutation

Strong mutation gives an effective but costly result when considering the whole program. On the other hand, weak mutation reduces considerably the number of tests, but it is only effective for components of a program. Hence, *firm mutation* has been proposed to balance the computational cost and the acceptable result [KOO_96]. Firm mutation considers a component of the program, which is chosen by the user, such as: a loop, a procedure … Firm mutation compares the state of the program at a given moment. Both strong mutation and weak mutation are particular cases of firm mutation [OFFU_96].

1.3.3 Selective mutation

For a given program, the cost of mutation testing is essentially related to the number of generated mutants. In fact, each mutant should be executed at least on a test vector. Intuitively, one way to reduce this cost is to decrease the number of generated mutants. This approach is called *selective mutation* [MATH_91] [OFFU_93]. Selective mutation can be done in two ways: either by a random reduction of the mutant number (mutation-x%), or by a complete omission of operators that generate the most mutants.

The second approach seems more promising and really reduces the complexity of mutation testing. In fact, Offut et al. [OFFU_96] have done some experimentation on the MOTHRA tool. They classified the 22 initial mutation operators into three principle groups:

- Group "R": operand replacement, which replaces each operand in a program with another legal operand.
- Group "E": expression modification, which replaces operators and inserts new operators.
- Group "S": statement modification, which modifies the whole statement.

Using strong mutation as reference, they obtained some good result by combining two groups simultaneously. Also, considering the E group only (called E-selective mutation), it was possible to obtain a mutation score of 99.51% with a reduction of the mutant number of 77.56%.

Hence, selective mutation clearly shows well that the question of the set of non-selective mutation operators is still to be solved. This domain, however, has not been deeply explored, and it leaves researchers with multiple directions to follow.

1.4 Test generation based on mutation

In this section, we describe a technique for automatically generating test data. This technique is based on mutation analysis and has been specifically developed to generate test data that kill mutants for a program.

In mutation testing, the aim of the design is to create the test data that are relatively adequate with regard to mutation operators, i.e., test data give a mutation score of 100%. The design therefore will look for a set of test data that only contains the effective test vectors, i.e., that kill at least one mutant. One way to automatically generate test data that kill mutants is to eliminate the ineffective test vectors. Such a filter can be described by mathematical constraints. Hence, the test data generation becomes the solution of an algebraic problem. The principle is the following:

Suppose that the mutant is represented by only a single modification in the program, the state of the mutant must differ from that of the original program after the last execution of the mutated statement, otherwise both mutant and the original program produce the same output. This characteristic is called the necessity condition so that the test vector kills the mutant. However, this condition is not enough. In fact, for a test vector to kill a mutant, it must create incorrect output, i.e., the final state of the mutant differs from that of the original program. It is the *sufficiency condition*. Determining the sufficiency condition implies knowing in advance the path that a program will take, which is obviously a difficult problem to solve.

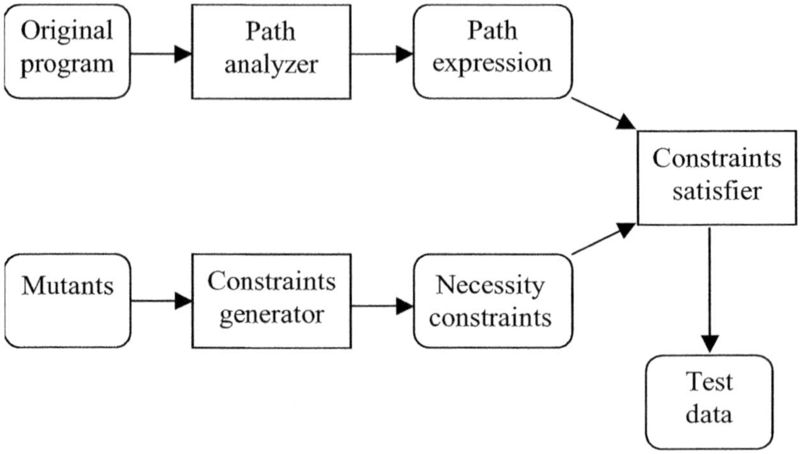

Figure 4.3-3. Test data generator based the constraints

Figure 4.3-3 shows the principle of a test data generator based on the constraints, proposed by DeMillo and Offut [DEMI_91]. The problem of test data generation, therefore, becomes a problem of constraint satisfaction. In fact, generated test data must satisfy, on the one hand, the necessity condition associated with mutants, and on the other hand, the expression accessibility of the mutated statements. The constraint satisfier then combines the expression accessibility and the necessity condition in order to establish the final constraint.

Finally, to generate test data that satisfies the final constraint, the satisfaction process proceeds in two phases: the symbolic execution and the test data generation. The symbolic execution transforms the final constraints into the primary inputs of the program. A random test data generation is carried out to attribute a value to each primary input of the program, which respects the final constraints. The authors, DeMillo and Offut, adopted an iterative heuristic to generate test data. It begins by randomly generating a value for the variable the domain of which is the smallest and then the final constraint is simplified with the aim of the chosen value. This process is repeated until all variables of the constraint have been assigned a value. This is an effective test data generation.

1.5 Functional testing method

1.5.1 Motivations

Increasing complexity of integrated circuits make very costly the test data generation based on low-level fault models. The progress in the CAD (Computer-Aided Design) field allows designers to describe circuits at a high level of abstraction: the function level. So, we propose a new approach of functional testing dedicated to the VHDL language. This approach addresses simultaneously the validation at functional level against design faults and the testing at hardware level, circuit implementation, against hardware faults. Its interest; therefore, is double.

1.5.2 Mutation testing for hardware

This approach consists in adapting the software testing method, mutation testing, to hardware in order to generate test data from VHDL functional descriptions of circuits. One aspect of this adaptation is to introduce a new functional fault model in the hardware domain.

1.5.2.1 Functional fault model

To validate VHDL descriptions, we have chosen the set of mutation operators shown in Table 4.3-1. This set constitutes, at the moment, a sub-set of mutation operators used for software testing, which is applicable on the sub-set of VHDL constructions.

Table 4.3-1. Mutation operators for VHDL

Type	Description
AOR	Arithmetic operator replacement
ABS	Absolute value insertion
CR	Constant replacement
CVR	Constant for variable replacement
LOR	Logical operator replacement
ROR	Relational operator replacement
NOR	No operation replacement
VCR	Variable for constant replacement
VR	Variable replacement
UOI	Unary operator insertion

The chosen set of mutation operators keeps all the criteria of software testing originally guaranteed by mutation testing. It guarantees the statement coverage, branch coverage, condition coverage, extreme value coverage, domain perturbation, etc. Satisfying these criteria at the functional level means that we exercise systematically the different functions constituting the program. This increases our confidence in the fact that the description does not contain design faults, and conforms to its specification. These criteria can be considered as the metrics that guide the methods intended from validation to design.

1.5.2.2 Enhancement process

Indeed, mutation testing considers VHDL operators with the same complexity degree of testing, then it generates a fixed number of test vectors for each one of the operators. The hardware implementation is, however, very sensitive to the nature and to the data type of their operands; i.e., the bit-width of the data buses.

To overcome these problems, Al-Hayek proposed to divide the operators into two categories: simple and complex. The *simple operators*, which produce simple mutants, do not have an impact on the hardware implementation and so they preserve their software characteristics. On the other hand, the *complex operators*, which produce complex mutants, influence the hardware implementation of the component. Generally, they are the operators that manipulate on the data buses. A test vector can kill a simple mutant, but it is necessary to have several test vectors to kill a complex mutant.

Al-Hayek considered that the AOR, LOR, ROR operators are complex and he studied theorically and experimentally to compute probabilistically the N necessary number of test data to test these complex operators. The principles of the probabilistic approach are:

Let $O(C_1, \ldots, C_D)$ be an complex operator composed of D identical cells which are independent of each other; $P_{1,1}$ the probability that a randomly generated test vector ensures a test data *d* on one cell input; $P_{N,D}$ the probability that N randomly generated test vectors ensure *d* on the input of each of the D cells; Al-Hayek proposed the following equation

$$N \approx \frac{\log\left(-\dfrac{\log P_{N,D}}{D}\right)}{\log(1-P_{1,1})}$$

$P_{N,D}$ is also the desired mutation score, for example $P_{N,D} = 95\%$.

$P_{1,1}$ is computed according to each of the complex operators: $P_{1,1} = 0.25$ for logical operators (LOR), $P_{1,1} = 0.125$ (optimistic approach) and

$P_{1,1} = 0.0625$ (pessimistic approach) for arithmetical (AOR) and relational operators (ROR).

To conclude, it is necessary to generate N test vectors in order to kill N times the mutant generated by a complex operator. The number N depends on the chosen approach and also on the operator type.

1.5.2.3 Testing for VHDL functional descriptions

The effective test data generation can be made either by a method derived form mutation like the generation guided by the constraints, or by any other mechanism of random/deterministic manual/automatic generation. The mutation approach is valuable and constitutes a tool to generate effective test data.

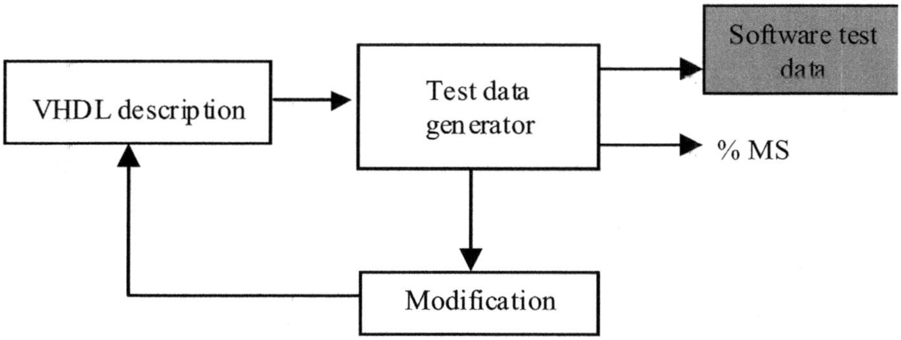

Figure 4.3-4. Mutation testing for VHDL descriptions

During the application of mutation testing, the VHDL description is simulated on the generated test, if the simulation result is not correct, then there is a (design) fault in the description.

1.5.2.4 Approach validation

Mutation testing, as presented, constitutes a systematic method for validating design. A VHDL description that passes a set of test data with a sufficient mutation score can confidently be considered as correct and in conformity with its specification.

However, in hardware testing, the most important thing is to eliminate hardware faults of electronic component implementation. We will measure the mutation testing effectiveness on this kind of fault. For this, we propose the architecture, called validation process, in Figure 4.3-5.

Remark: in the hardware testing literature, most testing strategies consider the gate-level description as the hardware description of the component. In the validation process, we need a logic synthesis tool to generate gate-level description from VHDL description.

To measure the effectiveness of the set of test data (generated by mutation approach), it is necessary to have a gate-level fault simulator. This simulator aims to compute the MFC (Mutation Fault Coverage) of hardware faults ensured by the set of test data.

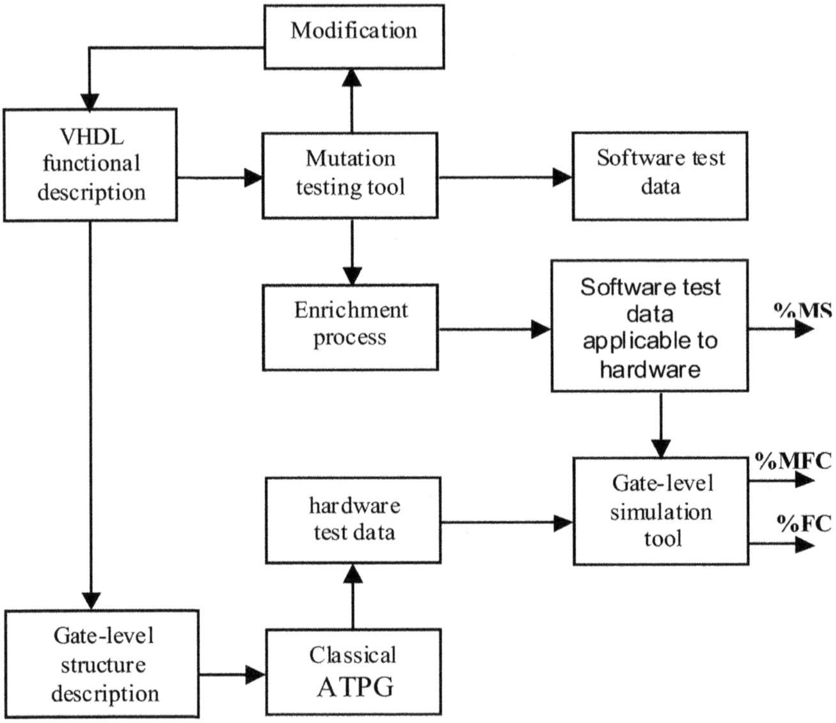

Figure 4.3-5. The validation scheme

1.5.2.5 Conclusion

Our method shows that the mutation technique, originally proposed for software testing, has been adapted for generating test data from VHDL functional descriptions of the electronic components, and thank to the enhancement process, this test data takes into account hardware characteristics in order to ensure a good coverage MFC of hardware faults.

2. THE ALIEN TOOL

This part presents the ALIEN tool -developed by the LCIS laboratory in Valence - which is a complete mutation testing environment for programs described in VHDL. Firstly, we will describe the tool, then we will present some experimental results by using it for a set of given VHDL programs.

2.1 The implementation tool

As we have seen, mutation testing allows a set of test data to be valued for a given program. The principle of mutation testing is to generate a defined set of versions of the program (called mutants). A mutant is generated by introducing in the original program only a single semantic modification as long as it is syntaxically correct. Mutation testing permits the effectiveness of a set of test data to be measured. This effectiveness depends on the capacity of the set of test data to distinguish the original program and its mutants.

The laboratory LCIS's ValSys team in Valence has adapted this method to hardware by applying it on functional descriptions and has shown that this kind of testing is effective simultaneously for design validation of a hardware system and also for logical fault detection.

Also, a mutant generator associated with a test data generator, which allows software or hardware products to be correctly tested in a less costly way, was developed by the ValSys team. This tool is called ALIEN.

2.1.1 General presentation of the tool

The ALIEN tool, as represented in Figure 4.3-6, allows three principle functions to be carried out as follows:

- the mutant generation: the mutant generator allows a mutation description table to be generated from an original program and a set of mutation operators, all of this relies on the required type of mutation.
- the test data generation: the test data generator allows a set of test data to be generated from the original program, the mutation description table, and if necessary an enhancement table. The test data generation will be done either in a random way or in a deterministic way. In the deterministic way, the test data generator uses a constraint generator and a solver, ECL*i*PS*e*, allowing the domains of possible inputs to be determined for test data. We also filter test data to increase their effectiveness and at least to avoid redundancy.
- the comparison process: the comparator executes the original program and its mutants with the same test data. It compares the results obtained,

and it kills all the mutant output which differs from that of the original program.

We will now describe more precisely the ALIEN tool by detailing these three functions above.

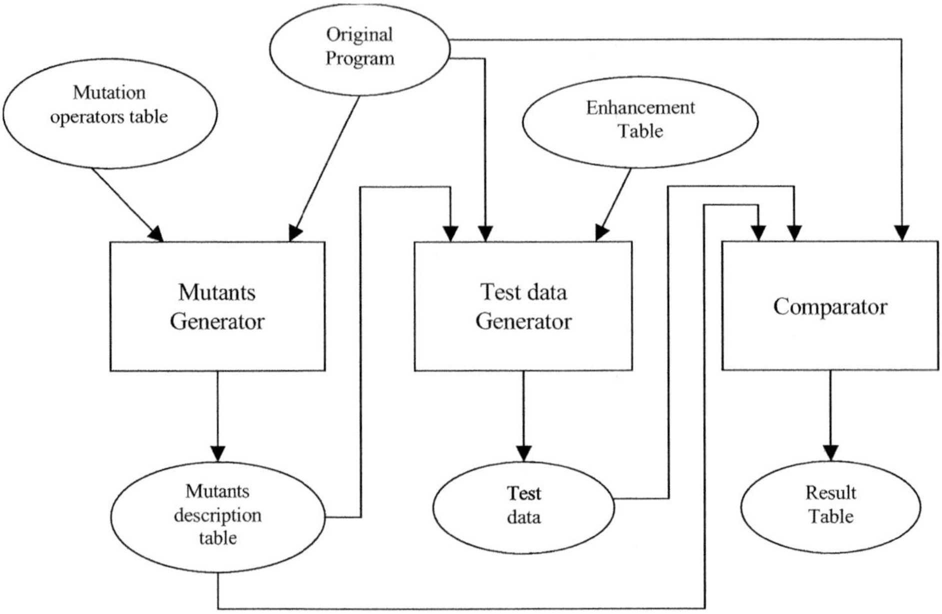

Figure 4.3-6. Data-flow diagram of the ALIEN tool

2.1.2 ALIEN detailed description

2.1.2.1 Mutant generator

The mutant generator is composed of different stages:

- Using FreeHDL software, we construct the syntactic tree of the VHDL original program and create a symbol table.
- Then, mutant descriptions are generated from the syntactic tree, modifiable mutation operators of the original program, and the chosen type of mutation. Each node of the syntactic tree is traversed, if this node can be mutated with respect to mutation operators in order to generate a mutant. All of these descriptions are stocked in a table, which is called mutant description table.

Remark: note that during mutant generation the user chooses the type of mutation that he (she) wants to realize. So, the generator can modify automatically the mutation operator table or act directly on mutants according to the type of mutation.

Hence, mutant generation is done automatically according to the chosen mutation and the user's choices.

2.1.2.2 Test data generator

Test data generation can be carried out randomly or deterministically. For the random generation, test data will be filtered by only keeping the test vectors that kill mutants. The deterministic generation is made by the intermediary ECL*i*PS*e*, which resolves the constraint system (composed of accessibility constraints and necessity constraints). Some examples of necessity constraints are given in Table 4.3-2. It is the test data generation based on constraints shown in the previous part.

Table 4.3-2. Examples of necessity constraints

Type	Original expression	Mutated expression	Constraint
AOR	Exp1 + Exp2 Exp1 - Exp2 Exp1 / Exp2 Exp1 * Exp2	Exp1 - Exp2 Exp1 + Exp2 Exp1 * Exp2 Exp1 / Exp2	Exp2 <>0 Exp2 <>1 Exp2 <> 1 and Exp2<>0
ABS	Abs(Exp) Negabs(Exp)	Negabs(Exp) Abs(Exp)	Exp<>0
CR	Const1	Const2	Const1 <> Const2
CVR	Const1	Var1	Const1<>Var1

Once the constraint system is obtained, we do the constraint resolution in two stages:

- symbolic execution: transforming final constraints into primary inputs of the program.
- test data generation: generating randomly test data in order to assign a value to each primary input of the program, which respects final constraints.

In parallel, in these stages, the generator can consider the enhancement tables as input. So, for the mutants generated by complex operators, it is necessary to generate N test vectors in order to kill N times each of these mutants, where N will be given by the enhancement tables.

Once these stages are finished, the results obtained, i.e., test data, will be stocked in a file (which by default has the format of the TESTGEN tool of SYNOPSYS)

2.1.2.3 Comparator

This last function allows us at first to generate a "testbench", which permits us to simulate the circuit described by the VHDL original program and one of its mutants on a test vector generated. For each mutant, such a program will be constructed automatically. This procedure can be described by the figure 4.3-7.

We notice that the outputs of the original program and of the mutant are connected one to one by "XOR". Hence, if the output is 1, then outputs of the original program and that of the mutant are different, the mutant is killed; otherwise if the output is 0, then outputs of the original program and that of the mutant are not different, the mutant is alive. It is therefore necessary to use another test vector and so on in succession.

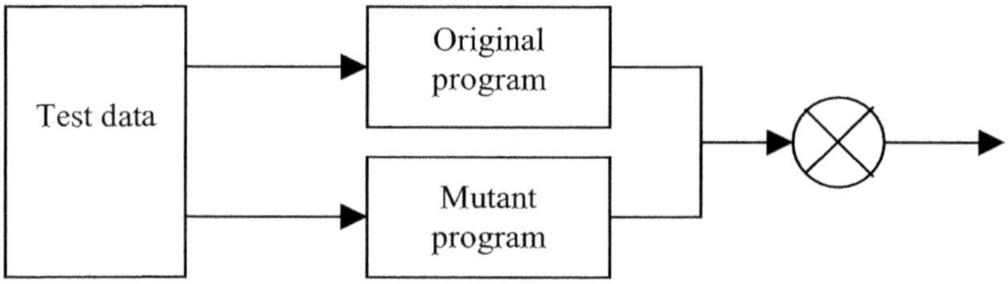

Figure 4.3-7. Modeling circuit serving for generating testbench

2.2 Experimental work

In our experimental work, we consider some circuits of different types and of different software and hardware complexity (concerning VHDL description). The set of circuits can be divided into two principle sub-sets: the circuits of "data path" type for data processing, and the circuits of "control" type (controller).

The "path data" sub-set contains combinational and sequential circuits of different software and hardware complexity:

- DEC: a decoder realizing several conditional statements fit, which makes complex its VHDL description with regard to software (combinational circuits).
- ALU: an arithmetic and logical unit, which constitutes an example type with an average software and hardware complexity (combinational circuits).
- EWF: an elliptical wave filter, which constitutes an example type (benchmark) for the high level synthesis (combinational circuits).

- DE: a differential equation, which also constitutes an example type for the high level synthesis (sequential circuits).

The circuit sub-set of control is composed of four controllers called b01, b02, b03 and fsm from the ITC'99 benchmarks. They are examples taken from the research on testing finite state machines described in VHDL.

For all the circuits, we have generated test data by applying mutation testing with a mutation score MS greater than 95%. This is considered as sufficient with regard to software and also allows us to establish confidently that the VHDL descriptions do not contain design faults.

To measure hardware fault coverage of mutation testing (MFC), we have used the SYDNOPSYS™ synthesis tool in order to generate a gate-level structure of circuit that we consider as hardware implementation. The hardware faults considered are the stuck-at faults and stuck-open faults. To measure the hardware coverage faults MFC, we have used the HILO™ system for simulating hardware faults.

2.2.1 Before enhancement of test data

The results presented in Table 4.3-3 correspond to hardware coverage faults (stuck-at fault) obtained by the test data generated by mutation approach without enhancement process.

Table 4.3-3. MFC(%) before enhancement

Circuit	Size	Surface	%MFC
DEC	35	98	100
ALU (16 bits)	89	1200	88.4
DE (16 bits)	73	2292	92
EWF (16 bits)	96	5700	96
b01	45	124	98.9
b02	19	134	98.3
b06	31	98	99.1
Fsm	26	122	98.8

The hardware faults considered by the HILO™ fault simulator are stuck-at faults in the gate-level structure of the circuit. The size of test vector (*Size*) and the size of hardware (*Surface*) are also reported; they are important parameters in test quality measure. *Size* is a principle factor for measuring the application time of a test vector and *surface* for measuring the hardware complexity of a circuit. *Surface* is given by the SYSOPSYS™ synthesis tool, it is an indicative measure of the necessary silicon surface implementing the circuit.

2.2.2 After enhancement of test data

Once the enhancement process is applied to the circuits, Table 4.3-4 shows the results of different typical sizes of data which have been obtained. The other circuits such as the decoder and controllers do not appear on the table because the enhancement process is not applied to them: they do not contain arithmetic or logical operators with data buses as inputs. To reduce at least the test data size, we only present the results obtained by the optimist approach. The results corresponding to the pessimist are certainly better in terms of the fault coverage, but the size of test data is bigger.

Table 4.3-4. MFC(%) after enhancement

	ALU			DE			EWF		
Dimension	Size	Surface	%MFC	Size	Surface	%MFC	Size	Surface	%MFC
4 bits	198	300	99.7	393	892	99	896	630	100
8 bits	211	600	99.3	443	1392	99.7	996	1860	100
16 bits	232	1200	99.6	503	2292	99.9	1176	5700	100
24 bits	242	1800	99.2	523	3192	99.2	1221	11700	100
32 bits	253	2400	99.5	553	4092	99.1	1296	17400	100

We note that the hardware fault coverage obtained after enhancement process is *good* and *stable* in spite of data dimension. However, the test data size increases with the dimension; this is obvious because the hardware complexity also increases. The results presented in the table only consider the stuck-at faults.

2.2.3 Comparison with the classical ATPGs

We compare between mutation testing, HILO and HITEC as shown in Figure 4.3-5. The results are presented in Table 4.3-5. For the combinational circuits, the three approaches are similarly effective. However, mutation testing generates test data of a bigger size than HILO, because mutation testing cannot optimize its test data by examining gate-level structure, which it takes into consideration, contrary to ATPG. For the sequential circuits, HILO has some real difficulties. On the other hand, both HITEC and mutation testing are very effective on this kind of circuit. However, we note that mutation testing is more effective on the controllers.

Table 4.3-5. Comparison with HILO and HITEC

	Mutation		HILO		HITEC	
	MFC (%)	Size	HILO-FC	Size	HITEC-FC	Size
DEC	100	35	100	17	100	34
ALU (16 bits)	99.6	232	99	33	100	98
DE (16 bits)	99.9	503	85	76	100	301
EWF (16 bits)	100	1,176	99	168	100	360
B01	98.9	45	4.2	17	98.6	96
B02	98.3	19	7.5	11	97.4	62
B06	99.1	31	3.5	15	95.5	68
Fsm	98.8	26	6.3	19	96.8	82

3. CONCLUSION

3.1 Approach robustness

3.1.1 Robustness with regard to the different hardware implementations

The same test data is used to estimate its effectiveness in terms of MFC when the circuit is implemented in different ways (Figure 4.3-5). This make us very confident in the mutation approach for generating test data that is able to detect the majority of the hardware faults even before knowing the details of the hardware implementation of the circuit.

To generate different hardware implementations of description VHDL, we used the options (optimisation time, surface and time-surface) of the SYNOPSYS synthesis tool. We note that the mutation approach ensures some stability (robustness) on different hardware implementations of circuits. The results obtained (Table 4.3-6) show that the test data generated by the mutation are effective in detecting hardware faults even before knowing the hardware implementation of the circuit.

Table 4.3-6. MFC(%) with different synthesis options

	LIB1			LIB2		
	time-surface	time	surface	time-surface	time	surface
DEC	100	100	100	100	100	100
ALU (16 bits)	99.6	99.9	99.2	99.1	99.5	98.9
DE (16 bits)	99.9	100	99.5	99.5	99.9	99.4
EWF (16 bits)	100	100	100	100	100	100
b01	99.1	99.3	99	98.9	99.1	99
b02	98	98.2	98.2	98.3	97.9	98
b06	98.9	98.6	98.7	99.1	99.3	99
Fsm	99	99.2	98.8	98.8	99	98.8

3.1.2 Robustness with regard to the different hardware fault models

We show the capacity of mutation testing of detecting types of fault other than stuck-at faults. In our experiments, we consider the stuck-open faults besides the stuck-at faults. Table 4.3-7 shows the result obtained. We can see that the mutation approach is also effective in detecting the stuck-at faults and stuck-open faults.

The experimental results given show the interest of mutation testing. This technique can be applied just after the specification phase. On the one hand, it allows the functional description of the circuit to be validated against design faults. The mutation score associated with the set of test data is a quality measure for this abstraction level. On the other hand, the test data generated by using the enhancement process has shown itself to be extremely useful for testing hardware implementation.

Table 4.3-7. MFC(%) on stuck-at and stuck-open faults

	DEC	ALU 16 bits	DE 16 bits	EWF 16 bits	b01	b02	b06	fsm
MFC (%)	100	99.7	99.9	100	99	97.9	99.4	99

3.2 Limitations and Reusability

We have proposed a set of mutation operators for VHDL that at present constitute the functional fault model. This model guarantees a set of standard criterion of high level testing such as statement coverage, branch coverage, condition coverage and extreme value coverage. This allows mutation testing to be considered, at a functional level, as a systematic method of design validation process. Note that up to now this process has been done in the ad hoc way and often manually. The mutation score was proposed as a new metric for measuring the quality of the validation process.

The recent researches have shown encouraging results, although generalising the approach for other complex examples, particularly industrial circuits, needs more theorical and experimental work. For example, the set of mutation operators needs to be optimised and better adapted to hardware. The optimisation effort consists in determining a minimal set of mutation operators. We have shown before that the set used at present is not. This work is aimed at reducing at the most the complexity of mutation testing.

The set of mutation operators also needs to be better adapted to hardware by keeping the original characteristics of mutation testing. This consists in introducing new mutation operators for proper hardware faults. These faults can be the design faults or physical faults.

At present, the mutation approach has been considered as a unified testing method for functional and gate-level integrated circuits. This is the biggest originality of mutation testing.

A perspective of this work is the establishment of a generalised approach that foresees a single and unique method for testing an electronic system from specification to implementation hardware on silicon.

An interesting project will be the establishment of a test data generation method from the most abstract description of the system that can be repeated for each phase in the design cycle until the system is actually implemented on silicon. This approach supposes that the establishment of the enhancement process from one level to another one is possible. However, this process was not proved in a general way.

This approach is very interesting and encouraging, it allows test data to be reused (*reliability* notion). The reliability notion is much needed for all testing fields. It permits the test effort to be economized and test data to be effectively generated by taking into account information from all higher abstraction levels.

ACKNOWLEDGMENT

The author would like to thank T.B. Nguyen and M. Delaunay for their invaluable contribution to the design of the ALIEN tool.

Chapter 4.4

NEW ACCELERATION TECHNIQUES FOR SIMULATION-BASED FAULT-INJECTION

Fulvio Corno[1], Luis Entrena[2], Celia Lopez[2], Matteo Sonza Reorda[1] and Giovanni Squillero[1]

[1] *Politecnico di Torino, Dipartimento di Automatica e Informatica, Torino, Italy*
[2] *Universitad Carlos III, Area de Tecnologia Electronica, Madrid, Spain*

Abstract: Several important application sectors increasingly require the design of fault-tolerant circuits; reducing the cost and design time asks for a new generation of CAD tools, able to efficiently validate the adopted fault-tolerant mechanisms. This paper outlines a fault-injection platform supporting the injection of transient faults in VHDL descriptions. New techniques are proposed to speed-up fault-injection campaigns without any loss in terms of gathered information. Experimental results are provided, showing that the proposed techniques are able to reduce the total time required by fault-injection campaigns by at least one order of magnitude.

Key words: Fault-Injection, VHDL, Fault Tolerance

1. INTRODUCTION

The growing application of electronic systems in safety-critical applications asks for new techniques to be applied in the design of fault tolerant circuits. At the same time, the cost and time-to-market constraints strongly affect the design of these circuits, and suitable techniques and tools are continuously needed to face these constraints. Finally, the adoption of

A. Benso and P. Prinetto (eds.), Fault Injection Techniques and Tools for Embedded Systems Reliability Evaluation, 41-48.

new sub-micron technologies also asks for effective techniques for reaching the required level of reliability.

In this framework, a major issue concerns the evaluation of the reliability of designed circuits, and fault injection emerged as an effective solution [NIKO_99]. When dealing with in-house designed ASICs or FPGAs, simulated fault injection (e.g., [JENN_94] and [DELO_96]) is normally more suitable than other approaches, such as those based on hardware fault injection (e.g., [ARLA_90] and [KARL_94]). This is due to several reasons:

- First, simulated fault injection provides a much higher flexibility in terms of supported fault models.
- Secondly, it supports reliability assessment at different stages in the design process, well before the stage in which a prototype is available.
- Finally, simulated fault injection can normally be more easily integrated into existing design flows.

On the other side, simulated fault injection may pose unacceptable CPU time requirements, being based on the simulation of the circuit, both in its fault-free version and in the presence of the enormous number of possible faults.

Several techniques have been proposed in the past to implement simulation-based fault-injection campaigns for transient faults:

- A first approach [JENN_94], [DELO_96] is based on modifying the system description, so that faults can be injected where and when desired, and their effects observed, both inside and on the outputs of the system. The main advantage of this method lies in its complete independence on the adopted simulator; however, it normally shows very low performance, due to the high cost for modification and recompilation for every fault.
- A second approach [GILM_99] relies on the simulation command language, or on the procedural interface provided by some simulators. The main advantage of this approach lies in the relatively low cost for its implementation, while the obtained performance is normally intermediate between those of the first and third approaches.
- A third approach deeply interacts with the simulation kernel and data structures to support the injection and observation features. This approach normally provides the best performance, but it can only be adopted when the code of a simulation tool is available and easily modifiable.

In this paper we outline some techniques for speeding-up fault-injection campaigns by adopting a simulation-based approach belonging to the second category among the three listed above. The proposed techniques may be fruitfully used for assessing the correctness and effectiveness of the fault tolerance mechanisms which are normally implemented within the ASIC and FPGA designs developed for fault tolerant applications. Coherently with the current design practice adopted in many design centers, we assume that the considered circuits are described at the RT level using some hardware description language such as VHDL or Verilog. Since they can be applied on pre-synthesis RT-level descriptions, the techniques we propose are particularly suited to allow early reliability assessment, thus providing significant benefits in terms of avoiding unnecessary recycling in the design flow.

For the purpose of this paper, only single bit flip faults on memory elements will be considered. The motivation is that in synchronous designs with moderately slow clocks transient faults are usually relevant for memory elements, only. Moreover, when designs are described using well-defined synthesizable description styles, memory elements may be deterministically recognized in the RT-level source; since gate-level optimization algorithms usually preserve memory elements, any gate-level bit flip faults on such memory elements can be modeled in a nearly exact way at the RT-level.

The basic idea behind our work is that speeding-up RT-level fault-injection campaigns may be obtained by following two main avenues of attack: first, clever techniques can be devised to generate and collapse the list of faults to be injected. Secondly, several optimization mechanisms can be defined to reduce the time required to simulate each fault.

A prototype of a complete fault-injection platform implementing the proposed techniques has been implemented and it has been evaluated on some benchmark circuits described in VHDL. Results are provided, showing the effects of the different techniques, and demonstrating that they are able to reduce the total time required by a fault-injection campaign by at least one order of magnitude. A previous version of this paper, reporting some preliminary experimental results on an industrial application, appeared in [BERR_02_A].

The paper is organized as follows: Section 2 introduces RT-level fault injections while the algorithm is presented in Section 3. Section 4 and 5 describe static fault collapsing, and Section 6 details the dynamic one. Section 7 presents some experimental results and Section 8 concludes the chapter.

2. RT-LEVEL FAULT-INJECTION CAMPAIGN

Single bit flip faults are generally termed *single error upsets* (SEUs). Let us denote SEU number *i* with S_i. $S_i = (ME_{L_i}, T_i^A)$, where L_i is the index of the fault location into the memory element list (*ME*), i.e., the memory element that changes its value due to the faults; and T_i^A is the fault activation time, i.e., the time instant when the fault location flips its value. We T_{sim} as the workload length, thus $\forall S_i \in \Psi : T_i^A \leq T_{sim}$.

Function C(*t*) represents the state of the fault-free circuit at workload instant *t*. The state takes into account all values produced on output ports and all values stored into memory elements: $\mathrm{C}(t) = \{\mathrm{PO}(t), \mathrm{ME}(t)\}$. $\mathrm{C}^{\mathrm{f}}_i(t)$ represents the same state under the effect of S_i. Clearly,

$$\forall t \in [0, T_i^A[: \mathrm{C}(t) = \mathrm{C}_i^{\mathrm{f}}(t)$$

since before T_i^A, $\mathrm{C}^{\mathrm{f}}_i(t)$ is not affected by S_i. The fault-free simulation is usually termed *golden run*.

It should be noted that the circuit is assumed to be synchronous sequential with, possibly, many clocks. By defining the quantum τ as the greatest common divider between all clocks periods, all significant time instants can be expressed in the form $a\tau$, with $a \in \mathbf{N}$. Functions $\mathrm{C}(\cdot)$ and $\mathrm{C}^{\mathrm{f}}_i(\cdot)$ are discrete and the golden run is a finite list of values $(\mathrm{C}(\tau), \mathrm{C}(2\tau), \ldots, \mathrm{C}(T_{sim}\tau))$, where $T_{sim}\tau$ is the length of the workload. To ease formulas, in the following $\tau = 1$.

The goal of the fault-injection campaign is to grade possible faults, by partitioning the set of all faults into four different sets:

- *Failure*: the set of all SEUs that, during the workload, produce a difference on an output port of the circuit.
- *Silent*: the set of all SEUs that, compared to fault-free circuit, never produce differences on output ports and, at the end of the workload, left no differences in memory elements.
- *Latent*: the set of all SEUs that, compared to the golden run, never produce differences on output ports, but, at the end of the workload, cause at least a memory element to differ.
- *Error*: the set of all SEUs that cause an error in VHDL simulation.

At the end of the fault-injection campaign, each SEU is classified in exactly one set.

In particular, faults belonging to the *Error* set represent a typical problem of high-level fault simulation, and they have no correspondent classification in gate-level campaigns. Simulation errors are caused, for instance, whenever a fault sets a signal to a value out of its declaration range. Once such an error has occurred, simulation may be halted. Indeed, several commercial VHDL simulators do halt simulation automatically in presence of such errors.

Next section details the fault-injection algorithm, while fault collapsing techniques are analyzed in Sections 4, 5 and 6. Fault-collapsing methodologies that can be applied only during the fault-injection experiment are called *dynamic*, while methodologies that can be applied before it are termed *static*. Static techniques are again classified in *workload-dependent* and *workload-independent*, whereas they require the analysis of the workload or not.

3. FAULT INJECTION

The purpose of fault-injection is to simulate each possible SEU comparing the behavior of the faulty circuit against the fault-free one and categorizing its effects.

In the adopted algorithm the fault-free circuit is simulated once. Then SEUs are simulated sequentially only for the time required for their categorization.

Checking the state of a simulation seeking differences is a resource-consuming task: interrupting and starting the simulation at each clock cycle comparing behaviors may introduce unacceptable overheads.

The adopted fault injection algorithm exploits simulator facilities. Circuit states are stored internally and compared via efficient built-in commands in specific moments.

Figure 4.5-1 sketches the proposed approach; a few additional considerations are made in following sections.

3.1 Checkpoints and Snapshot

Being sequential, the state of the circuit at time t is fully determined by the values stored into memory elements ME(t) and input stimulus PI(t). The state at a certain time t_o is called *checkpoint* and denoted with C(t_o). During a preliminary fault-free simulation, at times $t^g{}_1$, $t^g{}_2$, ..., $t^g{}_{Ng}$, checkpoints are stored to disk.

When injecting fault S_i, by definition no difference can exist between the behavior of the fault-free circuit and the faulty circuit before fault activation time T_i^A. Thus, it is possible to *load* the circuit state from the closest checkpoint C($t^g{}_j$) preceding fault activation time ($t^g{}_j < T_i^A$), instead of *simulating* circuit behavior from the beginning of the workload. Then, circuit is explicitly simulated until $T_i^A - 1$ and fault S_i injected as usual.

It should be noted that saving circuit states is a time- and space-consuming task. Thus it is necessarily to carefully trade-off the ability to resume simulation from *any* possible time instant with the amount of disk space required to save all these states and with the time required to store a checkpoint and to restart simulation from it.

In the current implementation, the N_g checkpoint instants are equally distributed over the workload.

Similarly, the fault-free circuit is fully simulated at the beginning of the fault injection campaign (the so-called *golden run*) and the trace of all memory elements and all internal signals is stored to a file called *snapshot*.

The snapshot is exploited by the simulator built-in commands to efficiently analyze the faulty behaviors of the circuit but, while conceptually analogous, cannot be used as a checkpoint to start simulation.

```
Simulate fault-free circuit and store checkpoints CP = ( C(t^g_1), C(t^g_2), ..., C(t^g_Ng) )
For each S_i {
    E_i = Ø                          // Empty equivalent set
    s = max_{1≤j≤N_g ∧ t^g_j ≤ T^A_i - 1} (j)   // Select the nearest state in golden run
    C^f_i(t_s^g) ← C(t_s^g)          // Load nearest circuit state from checkpoint list
    Simulate circuit until t = T^A_i - 1
    Inject S_i and calculate C^f_i(T^A_i)
    inc = 1                          // Initialize increment step
    T_end = T^A_i                    // Initialize interval
    while(S_i is not categorized) {
        T_start = T_end
        T_end = T_start + inc
        Simulate circuit calculating C^f_i(t), t ∈ ]T_start, T_end]
        Insert all faults S_j dynamically equivalent to S_i in set E_i
        If exists a fault S_j in set E_i already categorized {
            categorize S_i like S_j
        } else if Differences( C^f_i(t), C(t), t ∈ ]T_start, T_end] ) == Ø {
            categorize S_i as silent
        } else if exists one PO ∈ Differences( C^f_i(t), C(t), t ∈ ]T_start, T_end] ) {
            categorize S_i as error
        }
        inc = inc * 2
    }
    Categorize all faults S_j in E_i like S_i
}
```

Figure 4.4-1. Pseudo-Code of the Fault-Injection Procedure

3.2 Early stop

After injecting fault S_i, whenever differences propagate to an output port, the fault can be classified as *failure* and simulation may be halted.

Similarly, simulation can be stopped and fault classified as *silent* as soon all fault effects disappear. In fact, differences may be *generated* only at fault activation time T_i^A and then are merely *propagated*. More formally:

$$t_r \in]T_i^A, T_{sim}] \wedge C(t_r) = C_i^f(t_r) \Rightarrow \forall t \in [t_r, T_{SIM}] : C(t) = C_i^f(t)$$

being T_{sim} defined as the workload length as above.

Finally, a fault can be categorized as an Error as soon it causes a simulation error.

Generally, in three cases (*silent*, *failure*, *error*) SEUs are categorized before the end of the workload, while in one (*latent*), the fault-injection experiment is required to reach the end of the workload.

It has been experimentally observed that most of the SEUs can be classified either in the first clock cycles after injection or after a relatively long time.

Thus, in the proposed approach, after injecting fault S_i at T_i^A, the simulation is run for exponentially increasing amounts of time: one clock cycle after the activation of the SEU the faulty circuit signal traces are compared against the good ones; next comparison is performed after two clock cycles, then four, and so on.

This approach reduces simulation times, while minimizing the overhead required for starting and interrupting simulation.

3.3 Hyperactivity

It has been experimentally observed that some SEUs cause a great activity in memory elements, without producing a difference in any output. Such faults are called *hyperactive*, and adopting the above methodology would cause considerable performance degradation.

In the proposed approach, if a fault causes more than T_h differences from the fault-free circuit it is recognized as hyperactive and simulated for the whole workload without further checks. Only at the end, the full comparison is made.

3.4 Smart resume

Since starting the simulator is a time-consuming task, the fault simulator attempts to minimize this operation. If a fault S_i is categorized as *silent* at time t_s, clearly $\forall t \in [t_s, T_{SIM}] : C(t) = C_i^f(t)$, because all fault effects have disappeared from the circuit.

Thus, to inject a new fault j, if $T_j^A > t_s$ it is not necessary to restart the simulator from a checkpoint but it may be preferable to *simulate* the fault-free circuit from t_s to T_j^A - 1. In the current algorithm, after an early stop caused by a SEU classified as *silent*, the fault injector seeks for active faults that would be injected only between t_s and $t_s + T_{smart}$, where T_{smart} is a threshold that depends on circuit size and simulator startup time.

It is necessarily to carefully trade-off the ability not to restart the simulator with the amount of time required to uselessly simulate the fault-free circuit.

3.5 Dynamic Equivalencies

A set E_i of faults equivalent to S_i is dynamically built during simulation. The goal is to reduce the number of SEUs injected either by categorizing some faults before simulation, or stopping the current simulation as soon as the current SEU S_i is discovered equivalent to an already-simulated one. This optimization will be detailed later.

4. WORKLOAD INDEPENDENT FAULT COLLAPSING

This Section illustrates static fault-collapsing techniques that are uniquely based on the analysis of the circuit.

The study of the topology of the circuit helps determining the category of a fault. Analyzing an RT-level circuit it is usually easy to isolate memory elements from statements that will be synthesized to combinational logic. As a result, it is possible to infer a scheme of the circuit where only two types of elements appear: memory elements and combinational blocks (Figure 4.5-2).

Sets of memory elements are called *registers* and are denoted with *Ri* in the figure. Combinational blocks are denoted with *Ci*, primary inputs with *Ii* and primary outputs with *Oi*.

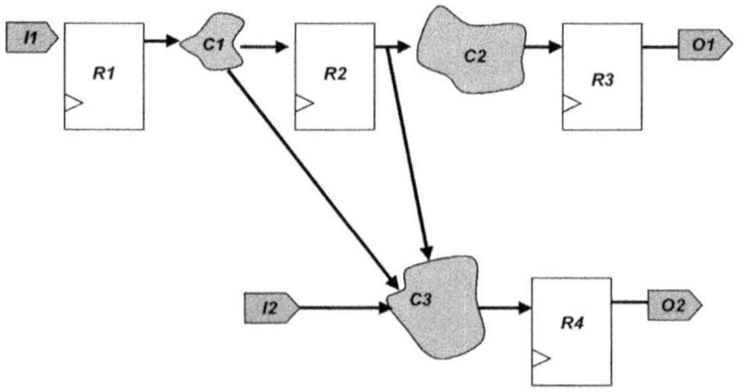

Figure 4.4-2. Circuit Topology

In this scheme it is possible to find the following cases:

- Primary input directly feeding a register

- Primary input going to a register through a combinational block
- Direct communication between two registers
- Communication between two registers through a combinational block
- Feedback in a register
- Direct communication from a register to a primary output
- Communication from a register to a primary output through a combinational block.

With the analysis of these elements it is possible to find *dominances* and *static equivalences* with respect to the faults to be injected. Dominant faults are those faults whose effect is the same as the effect for other faults and their simulation causes the same changes as the others but not the contrary.

On the other hand, equivalent faults are those faults whose effects are the same after a period of time, but the effects caused by any of them are not sub-set of the effects of the others.

Dominances and static equivalencies may reduce the size of the fault list, and speed-up the simulation process improving dynamic equivalencies, described later.

Direct registers to the outputs are very common in current designs, aero-spatial industry normally imposes this condition to all its designs in order to ensure Fault Tolerance in their equipments as indicated in the *ESA Guidelines*. All registers whose value is connected directly with the outputs of the circuit will have automatically all of their faults marked as failure. Moreover any register R_i whose output is connected directly and exclusively to another register R_j have its faults belonging to the same category as faults of the connected register R_j.

Further optimizations are possible, considering that all SEUs affecting registers with a given bit width can be categorized by analysing faults in a single memory element of the register, if the same operations and transformations are affecting all the bits. In addition, whenever the output of an internal counter is the overflow or a related function, each fault affecting it can be categorized by analyzing an equivalent SEU in the last memory element of the counter, although with a different activation time.

Experimental evidence (see Section 7) suggests that static test-bench dependent fault collapsing enables pruning about 5-10% of the total number of faults.

5. WORKLOAD DEPENDENT FAULT COLLAPSING

This Section illustrates static fault-collapsing techniques that are based on the simulation of the workload.

The fault-free circuit is simulated, and all read and write operations on memory elements are recorded. Each single read and write operation on each single bit is logged (the same granularity as SEUs). Figure 4.4-3 shows an example of read and write operations on bit 3 of a signal named sig.

Starting with the log, all possible SEUs are collapsed using the following rules:

- All SEUs between an operation (either read or write) and a write operation are marked as *silent*. Fault injection is useless, because their effect will be masked before any possible propagation. In the example, a SEU at $T^A = 9$ is silent.
- All SEUs between an operation (either read or write) and the subsequent read operation are equivalent. In the example, fault on `sig[3]` with $T^A = 1$ and $T^A = 2$ are equivalent, while the SEU at $T^A = 4$ is equivalent to the one at $T^A = 7$.

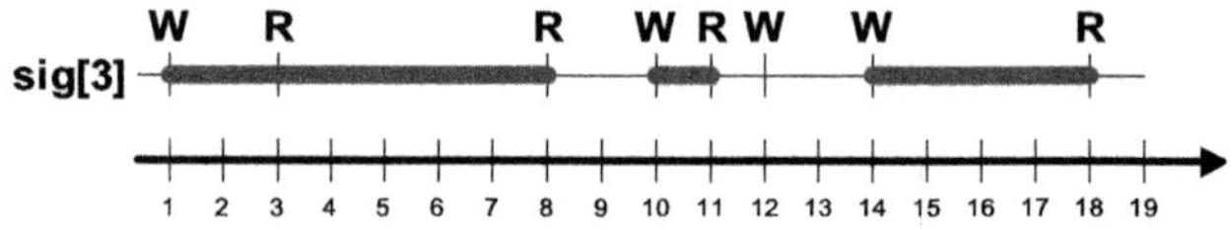

Figure 4.4-3. Memory element tracing

Experimental evidence suggests that workload dependent fault collapsing may enable pruning up to 80% of the total number of faults.

6. DYNAMIC FAULT COLLAPSING

This Section illustrates fault-collapsing techniques that may be activated during the fault-injection campaign. As anticipated before, this technique is exploited to discover equivalencies between faults. Fault equivalence is denoted with ≈.

During the simulation of SEU S_i, if at time $t_e > T_i^A$ differences between $C(t_e)$ and $C^f_i(t_e)$ are limited to the value of exactly one memory element ME_e, then S_i is equivalent to a bit-flip S_e on that memory element with $T_e^A = t_e$, i.e.,

$$S_j \approx S_e \text{ with } S_e = (FF_e, t_e)$$

Indeed, S_e may not be explicitly listed in the fault list, because static fault collapsing marked it equivalent to a different SEU S_f. In this case, for the transitive property, S_i will be marked equivalent to S_f

$$S_j \approx S_e \wedge S_e \approx S_f \Rightarrow S_j \approx S_f$$

As a result, during simulation the set E_i of SEUs equivalent to S_i is dynamically built. Whenever the fault injector is able to categorize S_i, all faults in E_i get the same classification.

It must also be noted that the newly discovered equivalent fault S_e may be already classified, even if $T_e^A > T_i^A$. First of all, there is no reason to presume that faults are injected in the same time order of their activation time (indeed, several optimizations are currently under study to optimize the order of injections). Moreover, fault S_e may be already classified because it has been found equivalent to a fault S_k with $T_e^A > T_i^A > T_k^A$. In this eventuality, fault S_i and all elements of E_i take the same classification as S_e.

Experimental evidence suggests that, exploiting dynamic fault collapsing, it is possible not to inject about 5% of the faults of the statically-collapsed fault list. Using a complete (not collapsed) list of SEUs, about 2 faults out of 3 may usually be classified without simulation.

7. EXPERIMENTAL RESULTS

A prototypical version of the fault-injection platform has been implemented in ANSI C, and consists of about 3,000 lines.

Circuit analysis exploits FTL Systems' *Tauri*™ parser, fault-list generation takes advantage of Synopsis VHDL Simulator, while the fault injector is currently based on *Modelsim*™ by Model Technology. Simulation states are saved using the *checkpoint* command, and subsequently loaded exploiting the *restore* option of the simulator. The faulty circuit and the golden run are compared taking advantage of the *waveform comparison* facilities built in the simulator.

The available prototype was tested on some ITC99 RT-level benchmarks; these benchmarks are representative of typical circuits, or circuit parts, that can be automatically synthesized as a whole with current tools and are described in [CORN_00].

The prototype was used to assess the reliability of b02, b14 and b15. Benchmark b02 is the smallest among ITC99 circuits; it consists of 70 VHDL lines synthesized to 28 gates and 5 memory elements. Benchmark b14 originally implemented a subset of the viper processor; it consists of 509 VHDL lines synthesized to approx 7K gates and 452 memory elements. Benchmark b15 originally implemented a subset of the 80386 processor; it consists of 671 VHDL lines synthesized to approx 13K gates and 449 memory elements. Since no workloads are available for ITC99 benchmarks, 2K clock-cycles random ones were used.

The total number of SEUs can be simply calculated, since failures may be injected on all memory elements in every clock cycle of the workload, excluding initialization. Table 4.4-1 reports benchmarks characteristics and the total number of SEUs, considering an initialization of 3 clock cycles.

Table 4.4-2 reports the estimated times required to run complete fault injection campaigns on a SPARC ULTRA Workstation with 256MB of RAM. These hypothetical fault injection campaigns consist in one fault free simulation (whose time is reported in the column [Golden run]) and one simulation for each possible SEU. Column [Avg SEU] reports the average time required to inject a single SEU, run simulation and compare the behavior of the circuit against the golden run. Comparisons were performed exploiting simulator waveform comparison mechanisms. The total amount of time required to run this hypothetical complete fault-injection experiment is reported in the last column [Total].

Table 4.4-1. Benchmarks characteristics

Benchmark	**VHDL**	**Memory Elements**	**Faults**
b02	70	5	9,985
b14	509	452	902,644
b15	671	449	896,653

Table 4.4-2. Estimated worst-case fault-injection campaign

Benchmark	**Golden run [s]**	**Avg SEU [s]**	**Total [s]**
b02	5.46	5.49	54,873
b14	7.11	6.28	5,670,966
b15	6.52	5.48	4,913,338

Static fault-collapsing mechanisms enabled a dramatic reduction in the fault list size. Table 4.4-3 reports the number of SEUs that require an explicit simulation [Active] against the original number [Total]. The table also details the effect of the two different static fault collapsing techniques: workload independent in column [IND] and dependent in column [DEP]. The former is slightly useful on small design, while enabled pruning up to 6% of the SEU on b15. The latter, on the other hand, is strictly dependent on the workload and the structure of the circuit.

Dynamic fault-collapsing mechanisms provide an additional reduction in the fault list size. Column [DYN] of Table 4.4-3 reports the number of SEUs that are found equivalent during fault injection campaign. This number is relatively small, since fault lists already underwent static fault collapsing. Last column [Red] reports the overall fault-list reduction.

Table 4.4-3. Static and dynamic fault collapsing

Bench	Total	Static		DYN	Active	Red [%]
		IND	DEP			
b02	9,985	41	1,049	494	8,895	10.92
b14	902,644	39,371	594,405	357	268,511	70.25
b15	896,653	52,775	700,210	501	143,167	84.03

Additionally, simulation optimizations reduce the time required for injecting faults. The first three columns of Table 4.4-4 [Single SEU] compare the average time required to run an un-optimized fault injection [UnOpt] against the average time required exploiting all techniques described above [Opt]. Column [Ratio] shows the ratio between optimized and un-optimized data.

Table 4.4-4. Overall comparison

Bench	Single SEU			Whole Campaign		
	UnOpt [s]	Opt [s]	Ratio	UnOpt [s]	Opt [s]	Ratio
b02	5.49	4.43	0.8069	54,873	35,374	0.6447
b14	6.28	1.59	0.2532	5,670,965	345,161	0.0609
b15	5.48	1.93	0.3522	4,913,338	272,768	0.0555

Finally, the last three columns of Table 4.4-4 [Whole Campaign] compare the average time required to run an un-optimized fault injection [UnOpt] against the average time required exploiting all techniques described above [Opt]. Column [Ratio] shows the ratio between optimized and un-optimized data.

Results in Table 4.4-4 show that by applying the techniques described in this paper it is possible to save up to 95% of the CPU time with respect to the plain approach to fault-injection campaigns without losing any information in terms of fault categorization. Results also suggest that the advantage stemming from the proposed techniques is greater when larger circuits are considered.

Results show that thanks to the significant reduction in the required CPU time, running 1-million-fault fault-injection campaigns may become a feasible task.

8. CONCLUSIONS

The possibility of performing massive fault-injection campaigns with acceptable CPU time requirements clearly improve the final quality and reliability of electronic circuits to be used in safety-critical applications. In this chapter we proposed a set of techniques able to significantly reduce the time required by fault-injection. Two approaches are adopted: from one side faults to be explicitly simulated are reduced by adopting suitable fault collapsing techniques. From the other, the average time required for the simulation of a single fault is also dramatically reduced. As a final result, fault-injection campaigns able to categorize the whole fault list on real-sized circuits are now feasible with commonly available processing facilities.

References

[AIDE_01] J. Aidemark, J. Vinter, P. Folkesson, J. Karlsson, "*GOOFI: Generic Object-Oriented Fault Injection Tool*", IEEE Int. Conf. on Dependable Systems and Networks, Göteborg, Sweden 2001, pp. 71-76

[AIDE_02] J. Aidemark, P. Folkesson, J. Karlsson, "*Path-Based Error Coverage Prediction*", Journal of Electronic Testing, Theory and Applications (JETTA), Vol. 16, June 2002, pp. 343-349

[ALHA_99] G. Al-Hayek, C. Robach, "*From validation to hardware testing: A unified Approach*", Journal Of Electronic Testing: Theory and Applications, Vol. 14, 1999, pp.133-140

[AMEN_96_A] A. M. Amendola, L. Impagliazzo, P. Marmo, G. Mongardi, G. Sartore,, "*Architecture and Safety Requirements of the ACC Railway Interlocking System*", 2nd International Computer Performance and Dependability Symposium, Urbana-Champaign, IL , USA, September 1996, pp. 21-29

[AMEN_96_B] A. Amendola, A. Benso, F. Corno, L. Impagliazzo, P. Marmo, P. Prinetto, M. Rebaudengo, M. Sonza Reorda, "*Fault Behaviour Observation of a Microprocessor System through a VHDL Simulation-Based Fault Injection Experiment*", IEEE European Design Automation Conference, Geneva (Switzerland), 1996, pp. 536-541

[AMEN_96_C] M. Amendola et al., "Experimental Evaluation of Computer-Based Railway Control Systems". Proceedings of FTCS-27, Seattle, June 1997

[AMEN_96_D] M. Amendola, et al., "Innovative techniques for analysis and experimental validation of signalling and automation systems". Proceedings of AEI-CIFE (in Italian), Firenze, September 1996

[AMEN_97] A. Amendola, L. Impagliazzo, P. Marmo, F. Pol,i., "*Experimental Evaluation of Computer-Based Railway Control Systems*", IEEE 27th Int. Symp. on Fault-Tolerant Computing (FTCS-27), Seattle, WA, USA, June 1997, pp. 380-384

[AMER_97] E.A. Amerasekera, F.N. Najm, "*Failure Mechanisms in Semiconductor Devices*", John Wiley & Sons, 1997

[ARLA_02] J. Arlat, J.C. Fabre, M. Rodríguez, F. Salles, "*Dependability of COTS Microkernel-Based Systems*", IEEE Transactions on Computers., Vol. 51, N. 2, February 2002, pp. 138-163

[ARLA_89] J. Arlat, Y. Crouzet, J. C. Laprie, "*Fault Injection for Dependability Validation of Fault-Tolerant Computing Systems*", IEEE 19th International Symposium on Fault-Tolerant Computing, 1989, pp. 348-355.

[ARLA_93] J. Arlat, A. Costes, Y. Crouzet, J. C. Laprie, D. Powell, "*Fault Injection and Dependability Evaluation of Fault-Tolerant Systems*", IEEE Transactions on Computers, Vol. 42, No. 8, August 1993, pp. 913-923

[ARLA_93] Jean Arlat, Nobuyasu Kanekawa, A. M. Amendola, Jean-Luis Dufour, Yuji Hirao, Joseph A. Profeta III, *"Dependability of Railway Control Systems"*, Panel at The 26th International Symposium on Fault-Tolerant Computing, Sendai, Japan. June 1996, pp. 150, 155

[ARLA_99] J. Arlat, J. Boué, Y. Crouzet, "Validation-based Development of Dependable Systems", IEEE Micro, Vol. 19, N. 4, July-August 1999, pp. 66-79

[ARMS_89] J.R. Armstrong, *"Chip-Level Modelling with VHDL"*, Prentice Hall, 1989

[ARMS_92] J.R. Armstrong, F.S. Lam, P.C. Ward, *"Test generation and Fault Simulation for Behavioural Models"*, Performance and Fault Modelling with VHDL, Englewood Cliffs, Prentice Hall 1992, pp. 240-303

[ASHE_01] P.J. Ashenden, *"The Designer's Guide to VHDL"*, 2nd Edition, San Francisco, CA, USA, Morgan Kaufmann Publishers, 2001

[AVRE_92] D. Avresky, J. Arlat, J. Laprie, Y. Crouzet, "*Fault Injection for the Formal Testing of Fault Tolerance*", IEEE 22th Annual international symposium on fault-tolerant Computing, Boston (MA), June 1992, pp. 345 – 354

[BARA_00] J.C. Baraza, J. Gracia, D. Gil, P.J. Gil, *"A Prototype of a VHDL-Based Fault Injection Tool"*, IEEE Int. Symposium on Defect and Fault Tolerance in VLSI Systems, Yamanashi (Japan), October 2000, pp. 396-404

[BARA_02] J.C. Baraza, J. Gracia, D. Gil, P.J. Gil, *"A Prototype of a VHDL-Based Fault Injection Tool. Description and Application"*, Journal of Systems Architecture, Vol. 47, N.10, 2002, pp. 847-867

[BAUM_01] R. Baumann, *"Soft Errors in Advanced Semiconductor Devices – Part I: The Three Radiation Sources"*, IEEE Trans. On Device and Material Reliability, 2001, pp. 17-22

[BENS_98_A] A. Benso, P. Prinetto, M. Rebaudengo, M. Sonza Reorda, "*EXFI: A Low-cost Fault Injection System for Embedded Microprocessor-Based Boards*", ACM Transactions On Design Automation of Electronic Systems, Vol. 3, No. 4, October 1998, pp. 626-634

[BENS_98_B] A. Benso, M. Rebaudengo, I. Impagliazzo, P. Marmo, *"Fault-List Collapsing for Fault Injection Experiments"*, Annual Reliability & Maintainability Symp., Annaheim, CA, USA, 1998, pp. 383-388

[BERG_02] I. Berger, *"Can You Trust Your Car?"*, IEEE Spectrum, April 2002, pp. 40-45

[BERN_02] K. Bernstein, *"High Speed CMOS Logic Responses to Radiation-Induced Upsets"*, Berkeley Univ. California, 2002, http://lithonet.eecs.berkeley.edu/variations/presentations/IBM_K.Bernstein_High%20Speed%20CMOS.pdf

[BERR_02_A] L. Berrojo, I. González, F. Corno, M. Sonza Reorda, G. Squillero, L. Entrena, C. Lopez, "*New Techniques for Speeding-up Fault-injection Campaigns*" IEEE Design Automation and Test in Europe Conference, Paris, France, March 2002, pp. 847-852

[BERR_02_B] L. Berrojo, F. Corno, L. Entrena, I. González, C. Lopez, M. Sonza Reorda, G. Squillero, "*An Industrial Environment for High-Level Fault-*

Tolerant Structures Insertion and Validation", IEEE 20th VLSI Test Symposium, Monterey, CA, USA, 2002, pp. 229-236

[BIND_75] D. Binder, E.C. Smith, A.B. Holman, *"Satellite anomalies from galactic cosmic rays"*, IEEE Trans. Nucl. Sci., 1975, vol. 22, pp. 2675

[BIND_98] M. Binderberger, "*Navy turns to off-the-shelf pcs to power ships (risks-19.75)*", RISKS Digest, Vol.19, N.76, May 1998

[BLAN_01] S. Blanc, J.C. Campelo, P. Gil, J.J. Serrano, *"Stratified Fault Injection using Hardware and Software-Implemented Tools"*, IEEE Design and Diagnostic of Electronic Circuits and Systems Workshop, Giör, Hungary, April 2001, pp. 259-266

[BLAN_02] S. Blanc, A. Ademaj, H. Sivencrona, J. Torin, P. Gil, *"Three Different Fault Injection Techniques Combined to Improve the Detection Efficiency for Time-Triggered Systems"*, IEEE Design and Diagnostic of Electronic Circuits and Systems Workshop, Brno, Czech Republic, April 2002, pp. 412-415

[BOEH_99] B. Boehm, "*Managing Software Productivity and Reuse*", IEEE Computer, Vol. 32, N. 9, September 1999, pp. 111-113

[BORK_99] S. Borkar, *"Design challenges of technology scaling"*, IEEE Micro, 1999, pp. 23-29

[BOUE_98] J. Boué, P. Pétillon, Y. Crouzet, *"MEFISTO-L: A VHDL-Based Fault Injection Tool for the Experimental Assessment of Fault Tolerance"*, IEEE 28th International Symposyum on Fault Tolerant Computing, Munich, Germany, 1998, pp. 168 173

[BREZ_01] E.A. Brez, *"By-Wire Cars Turn the Corner"*, IEEE Spectrum, 2001, pp. 68-73

[BUDD_78] T.A. Budd, R. DeMillo, R.J. Lipton, F.G. Sayward, "*The design of a prototype Mutation System for program testing*," ACM Nat. Computing Conference, 1978, pp. 623-627

[BUDD_81] T.A Budd, "*Mutation Analysis: Ideas Examples Problems and Prospects*", Computer Program Testing, 1981, pp. 129-148

[BURN_97] A. Burns, A.J. Wellings, "*Real-time Systems and their Programming Languages*", Addison Wesley, 1997

[CAIG_01] F. Caignet, S. Delmas-Bendhia, E. Sicard, *"The challenge of signal integrity on deep-submicron CMOS technology"*, IEEE proceedings, Vol. 89, N. 4, 2001

[CAMU_94] P. Camurati, F. Corno, P. Prinetto, C. Bayol, B. Soulas, "System-Level Modeling and Verification: A Comprehensive Design Methodology", 1st European Design and Test Conf., Paris, France, 1994, pp. 636-640

[CARR_98] J. Carreira, H. Madeira, J.G. Silva, "*Xception: A Technique for the Experimental Evaluation of Dependability in Modern Computers*", IEEE Transactions on Software Engineering, Vol. 24, N. 2, February 1998, pp. 125-136

[CARR_99_A] J.V. Carreira, D. Costa, J.G. Silva, "*Fault injection spot-checks computer system dependability*", IEEE Spectrum, Vol. 36, N. 8, August 1999, pp. 50-55

[CARR_99_B] J. Carreira, H. Madeira, J. Silva, "*Xception: A Technique for the Experimental Evaluation of Dependability in Modern Computers*", IEEE Transactions on Software Engineering, Vol. 24, N. 2, Feb. 1998, pp. 125-136

[CATA_01] A. Cataldo, *"SRAM soft errors cause hard network problems"*, EE Times, August 2001

[CATA_99] A. Cataldo, *"Intel scans for soft errors in processor designs"*, EE Times, June 1999

[CHEN_01] D. Chen, A. Messer, *"JVM Susceptibility to Memory Errors",* USENIX Java Virtual Machine Research and Technology Symposium, April 2001

[CHEV_01] P. Chevochot, I. Puaut, "*Experimental Evaluation of the Fail-Silent Behavior of a Distributed Real-Time Run-Time Support Built from COTS Components*", IEEE International Conference on Dependable Systems and Networks, Göteborg, Sweden, 2001, pp. 304-313

[CHSY_96] *Chorus/ClassiX r3 - Technical Overview*, Chorus Systems, Technical Report no. CS/TR 96-119.8, 1996, [www.sun.com/software/chorusos/overview.html]

[CLAR_95] J. A. Clark, D. K. Pradhan, "*Fault Injection A Method For Validating Computer-System Dependability*", in IEEE Computer, Vol. 28, No. 6, June 1995, pp. 47-56.

[CONS_02] C. Constantinescu, *"Impact of Deep Submicron Technology on Dependability of VLSI Circuits"*, Int. Conference on Dependable Systems and Networks, Washington, DC (USA), 2002, pp. 205-209

[CORN_00] F. Corno, M. Sonza Reorda, G. Squillero, "*RT-Level ITC 99 Benchmarks and First ATPG Results*," IEEE Design & Test of Computers, July-August 2000, pp. 44-53

[CORN_97] F. Corno, M. Sonza Reorda, G. Squillero, *"RT-Level ITC 99 Benchmarks and First ATPG Results"*, IEEE Design and Test of Computers, July-August 2000, pp. 44-53

[COST_00] D. Costa, T. Rilho, H. Madeira, "*Joint Evaluation of Performance and Robustness of a COTS DBMS Through Fault-Injection*", Dependable Systems and Networks Conference, New York, USA, June 2000, pp. 251-260

[CUNH_99] J.C. Cunha, M.Z. Rela, J.G. Silva, "*Can Software Implemented Fault-Injection be Used on Real-Time Systems?*", 3rd European Dependable Computing Conference, Prague, Czech Republic, 1999, pp. 209-226

[CZEC_93] E. Czeck, "*Estimates of the Abilities of Software-Implemented Fault Injection to Represent Gate-Level Faults*". Int. Work. on Fault and Error Injection for Dependability Validation of Computer Systems, Gothemburg, Sweden, 1993

[DAWS_96] S. Dawson, F. Jahanian, T. Mitton, T.L. Tung, "*Testing of Fault-Tolerant and Real-Time Distributed Systems via protocol Fault Injection*", IEEE 26th Int. Symposium on Fault-Tolerant Computing (FTCS-26), Sendai, Japan, 1996, pp. 404-414

[DELO_96] T.A. DeLong, B.W. Johnson, J.A. Profeta III, *"A Fault Injection Technique for VHDL Behavioral-Level Models"*, IEEE Design and Test of Computers, Vol 13, N. 4, Winter 1996, pp. 24-33

[DELO_96] T.A. Delong, B.W. Johnson, J.A. Profeta III, "*A Fault Injection Technique for VHDL Behavioral-Level Models*", IEEE Design & Test of Computers, Winter 1996, pp. 24-33

[DEMI_91] R. DeMillo, A. Offutt, "*Constraint-Based Automatic Test Data Generation*", IEEE Transactions on Computers, Vol. 17, N. 9, 1991, pp 900-910

[DURA_02] J. Durães, H. Madeira, "*Emulation of Software Faults by Selective Mutations at Machine-code Level*", IEEE 13th Int. Symposium on Software Reliability Engineering, Annapolis, MD, USA 2002

[ELEC_A] "Railway Application: Software for Railway Control and Protection Systems"

[ELEC_B] "Railway Application: Safety Related Railway Control and Protection Systems"

[EN50126] EN 50126:1999, *"Railway applications - The specification and demonstration of Reliability, Availability, Maintainability and Safety (RAMS)"* – Identical to IEC 62278 Committee Draft for vote 2002-05-11

[FASS_90] FASST Project Consortium. "*FASST: Fault Tolerant Architecture with Stable Storage Technology*". FASST Project (ESPRIT P5212) Technical Annex. 1990

[FOLK_98] P. Folkesson, S. Svensson, J. Karlsson, "*A Comparison of Simulation Based and Scan Chain Implemented Fault Injection*", IEEE 28th Int. Symp. on Fault-Tolerant Computing (FTCS-28), Munich, Germany, June 1998, pp. 284-293

[FUCH_96] E. Fuchs, "*An Evaluation of the Error Detection Mechanisms in MARS using Software Implemented Fault Injection*", IEEE 2nd European Dependable Computing Conference (EDCC-2), Taormina, Italy, October 1996, pp. 73-90

[FUCH_98] E. Fuchs, "*Validating the Fail-Silent Assumption of the MARS architecture*", 6th Dependable Computing for Critical Applications, Garmisch-Partenkirchen, Germany, 1998, pp. 225-247

[FULL_00] E. Fuller, M. Caffrey, A. Salazar, C. Carmichael, J. Fabula, "*Radiation Characterization and SEU Mitigation of the Virtex FPGA for Spaced Based Reconfigurable Computing*", NSREC, Reno, Nevada, USA, July 2000

[GAIS_02] J. Gaisler, "A Portable and Fault-Tolerant Microprocessor based on SPARC V8 Architecture", Int. Conf. on Dependable Systems and Networks, Washington, DC, USA, 2002, pp. 409-415

[GHOS_91] S. Ghosh, T.J. Chakraborty, *"On Behavior Fault Modeling for Digital Design"*, J. Electronic Testing: Theory and Applications, N. 2, 1991, pp. 135-151

[GHOS_99] A.K. Ghosh, M. Schmid, and F. Hill, "*Wrapping Windows NT Software for Robustness*", IEEE 29th annual International Symposium on Fault-Tolerant Computing, Madison (WI), June 1999, pp. 344-347

[GIL_92] P.J. Gil, *"Sistema Tolerante a Fallos con Procesador de Guardia: Validación mediante Inyección Física de Fallos"*, Tesis Doctoral, Departamento de Ingeniería de Sistemas, Computadores y Automática (DISCA), Universidad Politécnica de Valencia (Spain), September 1992

[GILB_97] P.J. Gil, J.C. Baraza, D. Gil, J.J. Serrano, *"High Speed Fault Injector for Safety Validation of Industrial Machinery"*, 8th European Workshop on Dependable Computing, Göteborg, Sweden, April 1997, pp.

[GILB_98] D. Gil, J. V. Busquets, J. C. Baraza, P. Gil, *"Using VHDL in the techniques of fault injection based on simulation"*, XIII Design of Circuits and Integrated Systems Conference, Madrid (Spain), 1998

[GILG_00] D. Gil, J. Gracia, J.C. Baraza, and P.J. Gil, *"A Study of the Effects of Transient Fault Injection into the VHDL Model of a Fault-Tolerant Microcomputer System"*, 6th IEEE International On-Line Testing Workshop (IOLTW 2000), Palma de Mallorca (Spain), July 2000, pp. 73-79

[GILM_99] D. Gil, R. Martinez, J.V. Busquets, J.C. Baraza, P. J. Gil, "*Fault Injection into VHDL Models: Experimental Validation of a Fault Tolerant Microcomputer System*", 3rd European Dependable Computing Conference, Prague, Czech Republic, September 1999, pp. 191-208

[GOOD_89] M. Goodman, A. McAuley, *"Exploiting the inherent Fault Tolerance of Asynchronous Arrays"*, Int. Conference on Systolic Arrays, Ireland, 1989, pp. 567-576

[GRAC_01_A] J. Gracia, J.C. Baraza, D. Gil, P.J. Gil, "*A Study of the Experimental Validation of Fault-Tolerant Systems using different VHDL-Based Fault Injection Techniques*", 7th IEEE International On-Line Testing Workshop (IOLTW 2001). Giardini Naxos, Taormina, Italy July 2001

[GRAC_01_B] J. Gracia, J.C. Baraza, D. Gil, P.J. Gil, "Comparison and Application of different VHDL-Based Fault Injection Techniques," Int. Symp. on Defect anf Fault Tolerance in VLSI Systems, San Francisco, CA, USA 2001, pp. 233-241

[GRAH_02] J. Graham, *"Soft errors a problem as SRAM geometries shrink"*, EBN, June 2002, http://www.ebnews.com/story/OEG20020128S0079

[GUEN_79] C.S. Guenzer, E. A. Wolicki, R.G. Allas, *"Single event upsets of dynamic RAM's by neutrons and protons"*, IEEE Trans. Nucl. Sci., Vol. NS-26, 1979, pp. 5048

[GUNN_89] U. Gunneflo, J. Karlsson, J. Torin, "*Evaluation of Error Detection Schemes Using Fault Injection by Heavy-ion Radiation*", IEEE 19th. International Symposium on Fault Tolerant Computing (FTCS-19), Chicago, MI, USA, June 1989, pp. 340-347

[GUTH_95] J. Guthoff, V. Sieh, "*Combining Software-Implemented and Simulation-Based Fault Injection into a Single Fault Injection Method*", IEEE 25th International Symposium on Fault-Tolerant Computing (FTCS-25), Pasadena, CA, June 1995, pp. 196-206.

[GWAN_92] S. C. Gwan, R. K. Iyer, "*FOCUS: An Experimental Environment for Fault Sensitivity Analysis*", IEEE Transactions on Computers, Vol. 41, No. 12, December 1992, pp. 1515-1526.

[HACH_93] A. Hachiga, K. Akita, Y. Hasegawa, *"The Design Concepts and Operational Results of Fault-Tolerant Computer Systems for the Shinkansen Train Control"*, 23th International Symposium on Fault-Tolerant Computing, Toulouse, France, June 1993, pp. 78-87

[HADJ_01] C.N. Hadjicostis, *"Coding Approaches to Fault Tolerance in Combinational and Dynamic Systems"*, Kluwer Academic Publishers, 2001

[HANR_95] S. Han, H. Rosenberg, K. Shin, "*DOCTOR: an Integrated Software Fault Injection Environment*", International Computer Performance and Dependability Symposium, Erlangen, Germany, April 1995, Springer-Verlag, pp. 204-213

[HARE_01] S. Hareland, J. Maiz, M. Alavi, K. Mistry, S. Walstra, C. Dai, *"Impact of CMOS process scaling and SOI on the soft error rates of logic processes"*, Symp. On VLSI Technology, 2001, pp. 73-74

[HAZU_00_A] P. Hazucha, C. Svensson, S. Wender, *"Cosmic-Ray Soft Error Rate Characterization of a Standard 0.6-um CMOS Process"*, IEEE Journal of Solid State Circuits, Vol. 35, N. 10, 2000, pp. 1422 -1429

[HAZU_00_B] P. Hazuscha, C. Svensson, *"Optimized Test Circuits for SER Characterization of a Manufacturing Process"*, IEEE Trans. Of Solid State Circuits, Vol. 35, N. 2, 2000, pp. 142-148

[HENN_93] Hennebert, G. Guiho, *"SACEM: A Fault Tolerant System for Train Speed Control"*, 23th International Symposium on Fault-Tolerant Computing, Toulouse, France, June 1993, pp. 624-628

[HILL_02] M. Hiller, A. Jhumka, N. Suri, "*On the Placement of Software Mechanisms for Detection of Data Errors*", IEEE International Conference on Dependable Systems and Networks, Washington, DC, USA, 2002, pp. 135-144

[HOFF_00] L. Hoffmann et al., "*Radiation Effects Testing of Programmable Logic Devices (PLDs)*", 2000 IEEE Nuclear and Space Radiation Effects Conference, Vancouver, B.C., Canada, July 2000,

[HOFM_00] G.J. Hofman, R.J.Peterson, C.J. Gelderloos, R.A. Ristinen, M.E. Nelson, A. Thompson, J.F. Ziegler, H. Mullfeld, *"Ligh-Hadron Induced SER and Scaling Relations for 16- and 64-Mb DRAMS"*, IEEE Trans. On Nuclear Science, Vol. 47, N. 2, 2000, pp. 403-407

[HOWA_01] J. Howard, *"Total Dose and Single Event Effects Testing of the Intel Pentium III and AMD K7 Microprocessors"*, MAPLD 2001

[HOWD_82] W.E. Howden, "*Weak Mutation Testing and Completeness of Program Test Sets*", IEEE Transaction on Software Engineering, Vol. SE-8, N. 4, 1982, pp. 162-169

[HSUE_97] M-C. Hsueh, T. K. Tsai, R. K. Iyer, "*Fault Injection Techniques and Tools*", IEEE Computer, Vol. 30, No. 4, April 1997, pp. 75-82.

[HWAN_00] S.H. Hwang and G. Choi, *"A Reliability Testing Environment for Off-the-Shelf Memory Subsystems"*, IEEE Design & Test of Computers, July-September 2000, pp. 116-124

[IYER_95] R.K. Iyer, *"Experimental Evaluations"*, IEEE Fault-Tolerant Computing Symposium Special Issue, Pasadena, CA, USA, June 1995, pp. 117-132

[JENN_94] E. Jenn, J. Arlat, M. Riemen, J. Ohlsson, J. Karlsson, "*Fault Injection into VHDL Models: The MEFISTO Tool*", IEEE 24th International Symposium on Fault Tolerant Computing (FTCS-24), Austin, Texas, USA, 1994, pp 66-75

[JOHN_00] A.H. Johnston, *"Scaling and Technology Issues for Soft Error Rates"*, 4th Annual Research Conference on Reliability, Stanford University, 2000

[JOHN_89] W. B. Johnson, "Design and Analysis of Fault Tolerance Digital Systems", Addison-Wesley Publishing Company, June, 1989

[KANA_93] G.A. Kanawati, N.A. Kanawati, J.A. Abraham, "*EMAX: An automatic Extractor of High-Level Error Models*", Computing Aerospace Conf., San Diego, CA, USA, 1993, pp.1297-1306

[KANA_95] G.A. Kanawati, N.A. Kanawati, J.A. Abraham, "*FERRARI: A Flexible Software Based Fault and Error Injection System*", IEEE Transactions on Computers., Vol. 44, N. 2, February 1995, pp. 248-260

[KANE_97] P.C. Kanellakis, A.A. Shvarstman, *"Fault-tolerant Parallel Computation"*, Kluwer Academic Publishers, 1997

[KAOI_93] W.L. Kao, R.K. Iyer, D. Tang, "*FINE: A Fault Injection and Monitoring Environment for Tracing the UNIX System Behavior under Faults*", IEEE Transactions on Software Engineering., Vol. 19, N. 11, November 1993, pp. 1105-1118

[KARL_94] J. Karlsson, P. Liden, P. Dahlgren, R. Johansson, U. Gunneflo, "*Using Heavy-Ion Radiation to Validate Fault-Handling Mechanisms*", IEEE Micro, Vol. 14, N.1, 1994, pp. 8-2

[KARL_95] J. Karlsson, P. Folkesson, J. Arlat, Y. Crouzet, G. Leber, J. Reisinger, *"Application of Three Physical Fault Injection Techniques to the Experimental Assessment of the MARS Architecture"*, 5th IFIP International Working Conf. Dependable Computing for Critical Applications, Champaign, IL, USA, September 1995, pp. 267–287

[KARL_98] J. Karlsson, P. Folkesson, J. Arlat, Y. Crouzet, G. Leber, J. Reisinger, "*Application of Three Physical Fault Injection Techniques to the Experimental Assessment of the MARS Architecture*", in Dependable Computing for Critical Applications (Proc. 5th IFIP Working Conf. on Dependable Computing for Critical Applications, Urbana IL, USA,

September 1995), (R. K. Iyer, M. Morganti, W.K. Fuchs, V. Gligor, Eds.), 1998, pp. 267-287

[KATZ_97] R. Katz, K. LaBel, J. Wang, B. Cronquist, R. Koga, S. Penzin, G. Swift, "*Radiation Effects on Current Field Programmable Technologies*", IEEE Transactions on Nuclear Science, Vol. 44, N. 6, December 1997, pp. 1945-1956

[KATZ_98] R. Katz, J. Wang, R. Koga, K. LaBel, J. McCollum, R. Brown, R. Reed, B. Cronquist, S. Crain, T. Scott, W. Paolini, B. Sin, "*Current Radiation Issues for Programmable Elements and Devices*", IEEE Transactions on Nuclear Science, Vol. 45, N. 6, December 1998, pp. 2600-2610

[KING_91] K.N. King, A. Jefferson Offutt, "*A Fortran Language System for Mutation based Software Testing*", Software-Practice and Experience, Vol. 21, 1991

[KING_93] K. King, A. Offutt, "*A Fortran Language System for Mutation-based Software Testing*", IEEE Design & Test of Computer, Vol. 10, N.3, 1993, pp. 16-28

[KOO_96] I. Koo, "*Mutation Testing and Three Variations*", http://www.geocities.com/Research Triangle/ Thinktank/5996/techpaps/mutate/mutation.html, 1996

[KOOP_99] P. Koopman, J. DeVale, "*Comparing the Robustness of POSIX Operating Systems*", IEEE 29th Int. Symp. on Fault-Tolerant Computing (FTCS-29), Madison, WI, USA, 1999, pp. 30-37

[LAPR_92] J. Laprie, "*Dependability: Basic Concepts and Terminology*", Dependable Computing and Fault-Tolerant Systems series, Vol. 5, Springer-Verlag, 1992

[LAPR_95] J.C. Laprie, *"Dependable Computing and Fault Tolerance Concepts and Terminology",* IEEE Computer, 1995, pp. 2-11

[LEEN_02] G. Leen, D. Heffernan, *"Expanding Automotive Electronic Systems",* IEEE Computer, Vol. 35, N. 1, January 2002, pp. 88-93

[LITT_93] B. Littlewood, L. Strigini, "*Validation of Ultra-high Dependability for Software-based Systems*", Communications of the ACM, Vol. 36, No. 11, November 1993, pp. 69-80

[LYNX_00] *LynxOS Real-Time Operating System*, LynuxWorks (formally Lynx RTS), 2000, [www.lynuxworks.com/products/index.html]

[MADE_02] H. Madeira, R.R. Some, F. Moreira, D. Costa, D. Rennels, "*Experimental evaluation of a COTS system for space applications*", International Conference on Dependable Systems and Networks, Bethesda, Maryland, USA, June 2002, pp. 325-330

[MADE_94] H. Madeira, M. Rela, F. Moreira, J. G. Silva, "*RIFLE: A General Purpose Pin-level Fault Injector*", 1st European Dependable Computing Conference, Berlin, Germany, 1994, Springer-Verlag, pp.199-216,

[MADR_01] "Fault Injection for the Time Triggered Architecture (FIT)", edited by C. Madritsch, Supplement of the International Conference on Dependable Systems and Networks, Special Track: European Dependability Initiative, Göteborg, Sweden, July 2001, pp d25-d27

[MAES_87] P. Maes, "*Concepts and Experiments in Computational Reflection*," ACM Conference on Object Oriented Programming, Systems, Languages and Applications, Orlando, FL, USA, 1987, pp. 147-155

[MART_99] R.J. Martínez, P.J. Gil, G. Martín, C. Pérez, J.J. Serrano, *"Experimental Validation of High-Speed Fault-Tolerant Systems Using Physical Fault Injection"*, 7th IFIP International Working Conference on Dependable Computing for Critical Applications, San Jose, California, USA, January 1999, pp 233-249

[MASS_96] L.W. Massengill, *"Cosmic and Terrestrial Single-Event Effects in Dynamic RAMs"*, IEEE Trans. On Nuclear Science, 1996, pp. 576 -593

[MATH_91] M.P. Mathur, "*Performance, effectiveness, and reliability issues in software testing*", IEEE 15th Annual International Computer Software and Application Conference, Tokyo, Japan 1991, pp. 604-605

[MAVI_00] D. Mavis, P. Eaton, *"SEU and SET Mitigation Techniques for FPGA Circuit and Configuration Bit Storage Design"*, MAPLD, September 2000

[MAY_ 79_A] T.C. May, M.H. Woods, *"Alpha-Particle-Induced Soft Errors in Dynamic Memories"*, IEEE Transactions on Electron Devices, 1979, pp. 2-9

[MAY_ 79_B] T.C. May, *"Soft Errors in VLSI: Present and Future"*, IEEE Trans. on Components,Hybrids, and Manufacturing Technology, CHMT-2, N. 4, 1979, pp. 377-387

[MESS_01] A. Messer , *"Susceptibility of Modern Systems and Software to Soft Errors"*, March 2001, http://www.hpl.hp.com/techreports/2001/HPL-2001-43.pdf

[MESS_82] C.G. Messenger, *"Collection of Charge on Junction Nodes from Ion Tracks"*, IEEE Transactions on Nuclear Science, Vol. 29, N. 6, 1982, pp. 2024

[MONG_93] G. Mongardi, "*Dependable Computing for Railway Control Systems*", 3rd IFIP International Working Conference on Dependable Computing for Critical Applications, 1993, Mondello, Italy, pp.255, 277

[MUSSE_01] O. Musseau, V. Cavrois, *"Silicon on Insulator Technologies: Radiation Effects"*, IEEE NSREC Short Course, 2001

[NGUY_ 98] D. Nguyen, C. Lee, and A. Johnston, *"Total ionizing dose effects on Flash memories"*, IEEE Radiation Effects Data Workshop, 1998, pp. 100-103

[NIKO_99] M. Nikoladis, "*Time Redundancy Based Soft-Error Tolerance to Rescue Nanometer Technologies*", IEEE 17th VLSI Test Symposium, April 1999, pp. 86-94

[NORM_96] E. Normand, *"Single Event Upset at Ground Level"*, IEEE Trans. Nucl. Sci., Vol. 43, 1996, pp. 2742 -2750

[OFFU_93] A. J. Offutt, G. Rothermel, C. Zapf, "*An Experimental Evaluation of Selective Mutation*", IEEE 15th International Conference on Software Engineering, Baltimore, Maryland, May 1993, pp. 100-107

[OFFU_96] J. Offutt, A. Lee, G. Rothermel, R. H. Untch, C. Zapf, "*An Experimental Determination of Sufficient Mutant Operators*", ACM Transaction on Software Engineering and Methodologu, Vol. 5, N. 2, 1996, pp. 99-118

[OHLS_98] M. Ohlsson, P. Dyreklev, K. Johansson, P. Alfke, *"Neutron Single Event Upset in SRAM-BASED FPGAs"*, IEEE Radiation Effects Data Workshop Record, Newport Beach, CA, USA, July 1998, pp. 177-180

[PALA_01] J.M. Palau, G. Hubert, K. Coulie, B. Sagnes, M.C. Calvet, S. Fourtine , *"Device Simulation Study of the SEU Sensitivity of SRAMs to Internal Ion Tracks Generated by Nuclear Reactions"*, IEEE Trans. Nucl. Sci,, Vol. 48, December 2001, pp. 225-230

[PARR_00] B. Parrotta, M. Rebaudengo, M. Sonza Reorda, M. Violante, *"Speeding-up Fault Injection Campaigns in VHDL Models"*, 19th International Conference on Computer Safety, Reliability and Security, Rotterdam, The Netherlands, 2000, pp. 27-36

[POST_03] PostgreSQL, [http://www.postgresql.org]

[PRAD_96] D.K. Pradhan, "Fault-Tolerant Computer System Design", Prentice-Hall 1996

[RIME_94] M. Rimén, I. Ohlsson, and J. Torin, "*On Microprocessor Error Behavior Modeling*", IEEE 24th Int. Symp. on Fault Tolerant Computing, Austin, Texas, USA, 1994, pp. 76-85

[RODR_00] M. Rodríguez, J.C. Fabre, J. Arlat, "*Formal Specification for Building Robust Real-time Microkernels*", IEEE 21st Real-Time Systems Symposium, Orlando, FL, USA, 2000, pp. 119-128

[RODR_02_A] M. Rodríguez, J.C. Fabre, J. Arlat, "*Wrapping Real-Time Systems from Temporal Logic Specifications*", 4th European Dependable Computing Conference, Toulouse, France, 2002, To appear - Available as LAAS Report no 02-121

[RODR_02_B] M. Rodríguez, J.C. Fabre, and J. Arlat, "*Assessment of Real-Time Systems by Fault Injection*", European Safety and Reliability Conference, Lyon, France, 2002, pp. 101-108

[RODR_02_C] M. Rodríguez, A. Albinet, and J. Arlat, "*MAFALDA-RT: A Tool for Dependability Assessment of Real Time Systems*", IEEE Int. Conf. on Dependable Systems and Networks, Washington, DC, USA, 2002, pp. 267-272

[RODR_99] M. Rodríguez, F. Salles, J.C.Fabre, and J. Arlat, "*MAFALDA: Microkernel Assessment by Fault Injection and Design Aid.*", Third European Dependable Computing Conference, Prague, Czech Republic, September 1999, pp. 143-160

[RONE_01] R. Ronen, A. Mendelson, K. Lai, Lu Shih-Lien, F. Pollack, J.P. Shen, *"Coming challenges in Microarchitecture and Architecture",* IEEE Proceedings, Vol. 89, N. 3, 2001, pp. 325-340

[SALL_99] F. Salles, M. Rodríguez, J.C. Fabre, J. Arlat, "*Metakernels and Fault Containment Wrappers*", IEEE 29th Int. Symp. on Fault-Tolerant Computing (FTCS-29), Madison, WI, USA, 1999, pp. 22-29

[SANT_01] N.D. Santos, D. Costa, "*eXception: An Evaluation Tool Towards the Demanding Availability of Networking Products"*, June 2001, FastAbstracts

[SCHE_00] L.Z. Scheick, G.M. Swift, S.M. Guertin, *"SEU Evaluation of SRAM Memories for Space Applications",* IEEE Trans. Nucl. Sci., 2000, pp. 61-63

[SEGA_88] Z. Segall, D. Vrsalovic, D. Siewiorek, D. Yaskin, J. Kownacki, J. Barton, R. Dancey, A. Robinson, T. Lin, "*FIAT - Fault Injection Based Automated Testing Environment*", IEEE 18th Int. Symp. on Fault-Tolerant Computing (FTCS-18), Tokyo, Japan, June 1988, pp. 102-107

[SEIF_01_A] N. Seifert, D. Moyer, N. Leland, R. Hokinson, *"Historical Trend in Alpha-Particle Induced Soft Errors Rates of the Alpha Microprocessor",* IEEE 39th Annual International Reliability Symposium, Orlando, Florida, 2001

[SEIF_01_B] N. Seifert , *"Frequency Dependence of Soft Error Rates for Sub-Micron CMOS Technologies",* IEEE International Electron Devices Meeting, 2001, pp. 323-326

[SHIV_02] P. Shivakumar, *"Modeling the Effect of Technology Trends on Soft Error Rate of Combinational Logic",* Int. Conf. on Dep. Systems and Networks, June 2002

[SIEH_97] V. Sieh, O. Tschäche, F. Balbach, "*VERIFY: Evaluation of Reliability Using VHDL-Models with Embedded Fault Descriptions*", IEEE 27th Int. Symp. on Fault-Tolerant Computing (FTCS-27), Seattle, WA, USA, June 1997, pp. 32-36

[SMIT_00] D. Smith, T. DeLong, B. W. Johnson, "*A Safety Assessment Methodology for Complex Safety Critical Hardware/Software Systems*", International Topical Meeting on Nuclear Plant Instrumentation, Controls, and Human-Machine Interface Technology, Washington, D.C., November 2000, 13 pages, (Invited Paper)

[SP109_94] "High-Performance I/O Bus Architecture: a Handbook for IEEE Futurebus+ Profile B", IEEE Standards Press. 1994

[SRIN_94] G.R. Srinivasan, H. K. Tang and P.C. Murley, *"Parameter-Free, Predictive Modeling of Single Event Upsets due to Protons, Neutrons, and Pions in Terrestrial Cosmic Rays",* IEEE Trans. Nucl. Sci., Vol. 41, 1994, pp. 2063-2070

[STOT_00] D. T. Stott, B. Floering, D. Burke, Z. Kalbarczyk, R. K. Iyer, "*NFTAPE: a framework for assessing dependability in distributed systems with lightweight fault injectors*", IEEE International Computer Performance and Dependability Symposium, March 2000, pp. 91-100

[SUEH_97] M. Sueh, T. Tsai, R.K. Iyer, *"Fault Injection Techniques and Tools",* IEEE Computer, Vol. 30, N. 4, April 1997, pp. 75-82

[SUNW_99] J. Sun, J. Wang, X. Yang, "*A Fault Injection Model and Its Application Algorithm for Testing and Evaluation of FTM*", Journal of Computer Research and Development of China, 1999

[VOAS_97] J. Voas, G. McGraw, L. Kassab, L. Voas, "*A 'Crystal Ball' for Software Liability*", IEEE Computer, Vol. 30, No. 6, June 1997, pp. 29-36

[VOAS_98] J.M. Voas, "*Certifying Off-the-Shelf Software Components*", Computer, Vol. 31, N. 6, June 1998, pp. 53-59

[VXWO_98] *VxWorks Realtime Kernel*, WindRiver Systems, 1998, [www.wrs.com/products/html/ vxwks52.html]

[WAKE_78] J. Wakerly, *"Error Detecting Codes, Self-Checking Circuits and Applications",* Elsevier North-Holland, 1978

[WANG_99] J. J. Wang, R. B. Katz, J. S. Sun, B. E. Cronquist, J. L. McCollum, T. M. Speers, W. C. Plants, *"SRAM Based Re-programmable FPGA for Space Applications",* IEEE Trans. Nucl. Sci., Vol. 46, N. 6, December 1999, pp. 1728-1735

[WROB_01] F. Wrobel, J.M. Palau, M.C. Calvet, O. Bersillon, H. Duarte, *"Simulation of Nucleon-Induced Nuclear Reactions in a Simplified SRAM Structure: Scaling Effects on SEU and MBU Cross Sections",* IEEE Trans. On Nuclear Science, Vol. 48, N. 6, 2001, pp. 1946-1952

[ZIEG_00] J.F. Ziegler, *"Trends in Electronic Reliability – Effects of Terrestrial Cosmic Rays",* 2000, http://www.srim.org/SER/SERTrends.htm

[ZIEG_79] J.F. Ziegler, W. A. Lanford, *"The effect of cosmic rays on computer memories",* Sci., Vol. 206, 1979, pp. 776

[ZIEG_96] J. Ziegler, H.W. Curtis, F.P. Muhlfeld, C.J. Montrose, B. Chin, M. Nicewicz, C.A. Russell, W.Y. Wang, L.B. Freeman, P. Hosier, L.E. LaFave, J.L. Walsh, J.M. Orro, G.J. Unger, J.M. Ross, T.J. O'Gorman, B. Messina, T.D. Sullivan, A.J. Sykes, H. Yourke, T.A. Enger, V. Tolat, T.S. Scott, A.H. Taber, R.J. Sussman, W.A. Klein, C.W. Wahaus, *"IBM experiments in soft fails in computer electronics (1978-1994)",* IBM Journal, Research and Development, Vol. 40, N. 1, Jan. 1996, pp. 3-16

[ZIEG_98] J. Ziegler, M.E. Nelson, J.D. Shell, R.J. Peterson, C.J. Gelderloos, H.P. Muhlfeld, C.J. Montrose,, *"Cosmic Ray Soft Error Rates of 16-Mb DRAM Memory Chips",* IEEE Journal of Solid State Circuits, Vol. 33, N. 2, February 1998, pp. 246-252

GPSR Compliance

The European Union's (EU) General Product Safety Regulation (GPSR) is a set of rules that requires consumer products to be safe and our obligations to ensure this.

If you have any concerns about our products, you can contact us on ProductSafety@springernature.com

In case Publisher is established outside the EU, the EU authorized representative is:

Springer Nature Customer Service Center GmbH
Europaplatz 3
69115 Heidelberg, Germany

Batch number: 10755292

Printed by Printforce, the Netherlands